AF733198

PHYSICAL TESTING OF PLASTICS— CORRELATION WITH END-USE PERFORMANCE

A symposium
sponsored by ASTM
Committee D-20 on
Plastics and
Subcommittee D20.10
on Mechanical Properties
AMERICAN SOCIETY FOR
TESTING AND MATERIALS
Philadelphia, Pa., 20 June 1979

ASTM SPECIAL TECHNICAL PUBLICATION 736
R. E. Evans, American Cyanamid Company,
editor

ASTM Publication Code Number (PCN)
04-73600-19

AMERICAN SOCIETY FOR TESTING AND MATERIALS
1916 Race Street, Philadelphia, Pa. 19103

Library of Congress Catalog Card Number: 80-69753
ISBN 0-8031-0768-4

NOTE

The Society is not responsible, as a body, for the statements and opinions advanced in this publication.

Printed in Baltimore, Md.
May 1981

Second Printing, Baltimore, Md.
April 1987

Foreword

The symposium on Physical Testing of Plastics—Correlation with End-Use Performance was held in Philadelphia, Pennsylvania, 20 June 1979. The symposium was sponsored by the American Society for Testing and Materials' through its Committee D-20 on Plastics and Subcommittee D20.10 on Mechanical Properties. R. E. Evans, American Cyanamid Company, presided as chairman of the symposium and editor of this publication.

Related ASTM Publications

Wear Tests for Plastics: Selection and Use, STP 701 (1980), 04-701000-19

Corrosion and Degradation of Implant Materials, STP 684 (1979), 04-684000-27

Disposal of Plastics with Minimum Environmental Impact, STP 533 (1980), 04-533000-19

A Note of Appreciation to Reviewers

This publication is made possible by the authors and, also, the unheralded efforts of the reviewers. This body of technical experts whose dedication, sacrifice of time and effort, and collective wisdom in reviewing the papers must be acknowledged. The quality level of ASTM publications is a direct function of their respected opinions. On behalf of ASTM we acknowledge with appreciation their contribution.

ASTM Committee on Publications

Editorial Staff

Contents

Introduction 1

A Rationale for Testing to Establish Suitability for an End-Use Application—
R. E. EVANS 3

Fracture Toughness Testing of HDPE Plastic Pipe—
P. FLUELER, D. R. ROBERTS, J. F. MANDELL, AND F. J. MCGARRY 15

Fatigue and Its Relationship to End-Use Performance—
H. S. LOVELESS, L. R. INGLE, AND C. E. GERRARD 32

Instrumented Impact Testing for Evaluating End-Use Performance—
D. R. IRELAND 45

Instrumented Impact Measurements on Some Polymers—
J. K. RIEKE 59

On Abrasion, Scratch, and Friction Testing—Relationship to End-Use Application—
J. F. CARROLL 77

Wear Testing—Correlation with End-Use Performance—
N. S. EISS, JR. 91

The Need for More Descriptive and Meaningful Physical Data on Plastic Materials—
F. S. PANG AND J. L. ISAACS 101

Testing Plastics for Performance: Meeting the Challenge of the Future—
J. L. O'TOOLE 109

Summary 114

Index 121

Introduction

The symposium on "Physical Testing of Plastics—Correlation with End-Use Performance" from which this volume was taken was organized to provide a forum for the discussion of the types of physical tests necessary to obtain data that are relevant to the intended application. Subcommittee D-20.10 on Mechanical Properties (of Plastics), under whose direction this symposium was organized, believed that such discussions were necessary because the concept of physical testing has been oversimplified in materials supplier's typical value tables. A major theme of the symposium and this volume is that physical properties needed for end-use applications must be relevant and cannot be presented as simple unchanging values that universally describe material behavior or component performance. Predictive testing rationale must consider both how properties change with time, temperature, and environment and how the boundary conditions and input excitations influence observed behavior. The evaluation of physical properties from the viewpoint of obtaining critical engineering data needs to be understood. This volume is a compilation of papers describing several types of procedures that exemplify how one analyzes the end-use requirements and correctly chooses tests and conditions.

Moreover, this volume is unique because emphasis is placed on the special rationale and philosophy required in making such measurements compared to testing employed for quality assurance, basic property studies and typical value data. Although the areas discussed are clearly not complete, the papers should describe for the reader the kind of considerations necessary to measure physical properties that govern the selection of any particular plastic for a specific application. The papers presented here have been successful both in illustrating various problems and in presenting some potential solutions. This book should be useful to both the physical testing experimentalist in developing meaningful physical data and the designer in requesting and selecting data relevant to the end-use application. Hopefully, it will serve to stimulate both groups and encourage closer liaison between the two disciplines. In addition, polymer chemists and physicists, materials scientists and engineers as well as applications engineers and any others who need to understand and apply test data should benefit from this volume. If the philosophy and methodology presented in this book are accepted, relevant data necessary to establish suitability of a material for a given end-use application will result. It is hoped that this volume will initiate the first step required for the adoption of an interdisciplinary approach to the development of guidelines necessary for the successful application of plastics as principal end-use materials for products in many different industries.

R. E. Evans
American Cyanamid Company, Stamford, Conn. 06904; symposium chairman and editor.

R. E. Evans[1]

A Rationale for Testing to Establish Suitability for an End-Use Application

REFERENCE: Evans, R. E., "**A Rationale for Testing to Establish Suitability for an End-Use Application,**" *Physical Testing of Plastics—Correlation with End-Use Performance, ASTM STP 736,* R. E. Evans, Ed., American Society for Testing and Materials, 1981, pp. 3–14.

ABSTRACT: The methodology necessary for selecting physical test procedures relevant to end-use applications is discussed. Emphasis is placed on the complex behavior of plastics and the fact that the physical properties of polymers cannot be presented as unchanging "typical values" that universally describe material behavior.

A basic rationale necessary to develop a sound philosophy of testing is first developed. An integrated approach of describing all physical tests simply as input/output measurements under given boundary conditions is presented. Three classes of test methods, that is, fundamental tests, component or practical tests, and hybrid tests are defined. The importance of avoiding hybrid types of tests is illustrated. An assessment of standard ASTM test methods is presented. Guidelines are established for selecting relevant physical testing procedures.

KEY WORDS: physical testing, evaluation, end-use performance, methodology, fundamental test, boundary conditions, environment, relevancy

The subject of this Special Technical Publication is physical testing and its correlation with end-use performance. In order to be able to obtain correlation with end-use performance, however, it is necessary to examine the basic rationale of physical testing. We must first understand why we test, and then the different philosophies involved in each area of testing. We must also understand what a physical test really is, the different types of physical tests that can be performed, and the influences of time, environment, and material history on the test results. The objective of this paper is to provide some guidelines for the methodology necessary for meaningful physical testing. The basic rationale necessary to develop a sound philosophy of testing is first developed. This is followed by an assessment of how to test and what to test for so that prime indicators of performance are obtained.

Reasons for Physical Testing

Reasons for testing include (1) quality assurance; (2) obtaining standard reference or typical value data; (3) research to obtain a fundamental under-

[1]Group leader, Chemical Research Division, American Cyanamid Co., Stamford, Conn. 06904.

standing of polymer structure; and (4) assessment of a material or article in relation to the function it is to fulfill, that is, establishment of the suitability for a given application. The difference between the latter case and the first three is one of philosophy as well as one of technique.

Quality assurance tests and tests used to obtain standard reference data require techniques that can be used either to assure process uniformity or to provide comparison data for materials under standard laboratory conditions. These tests are usually oversimplified and merely provide limited answers. Research of polymer structure requires techniques providing results that can be interpreted on a molecular scale. These tests, however, may not provide data in a form sufficiently close to that needed for relation to the actual use requirements of plastics. While they have a definite place in the testing picture, their limitations must be carefully considered.

Physical testing for end-use performance requires that the relevancy of the test be considered. The reason for testing in the latter case is to assess the behavior of a material or article in relation to the function it is to fulfill, in other words, to establish the material's suitability for a given purpose. Most simply, the purpose may be performance related, endurance related, or sensory related. More complex purposes may require a combination of any two categories or of all three categories. It is often helpful at the onset of an investigation to pinpoint which section of the interlocking circles shown in Fig. 1 that the application must satisfy.

Testing a plastic to determine suitability for a given purpose requires techniques that can be used to establish that a material or component fulfills its function by retaining all significant physical attributes throughout its expected service life. These techniques may be standard methods (modified when necessary to conform to the service condition) or may require development of a unique method. The techniques may be designed to measure intrinsic polymer properties, processing properties or product (article) properties [*1*][2].

A better appreciation of how meaningful physical test procedures can be established might be obtained if the following are first presented: (1) the scope of physical testing; (2) a classification system for the various types of tests; and (3) an assessment of standard ASTM tests.

Discretion must be used in selecting relevant test procedures because undoubtedly, too much physical testing is performed. Beatty and Sitz [*2*] estimate that 70 percent of the physical testing conducted is useless or ignored. The chief reason for this is either a failure to analyze the requirements necessary to solve the problem or a poor understanding of the limitations of the selected test methods. The accumulation of useless physical testing data is the result of poor planning or misunderstanding of the problem at hand and the proposed solution by the experimentalist. Such poor planning or

[2]The italic numbers in brackets refer to the list of references appended to this paper.

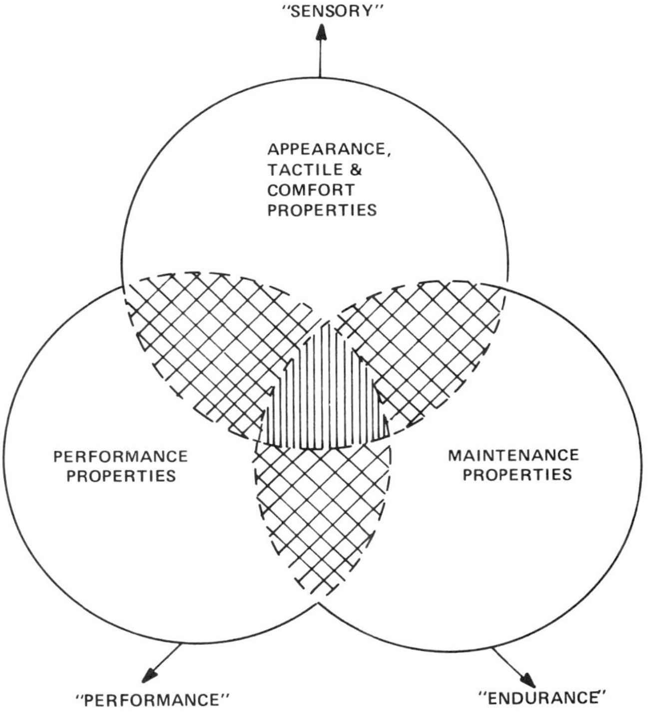

FIG. 1—*The suitability of a material or design for an end-use application is related to only its performance, endurance or sensory properties or some combination thereof (after Von Krevelen* [1]).

understanding can also result in not running the critical test, a pitfall more serious than obtaining too much data.

Scope of Physical Testing

The first rule of physical testing should be: physical properties of polymers cannot be presented as unchanging or "typical" values that describe a material's universal behavior. This fact is often overlooked by materials engineers.

One reason for the misuse of plastics' test procedures is that many of the tests (and testing equipment) used to characterize plastic materials were originally adopted with minor modifications from techniques used for the traditional materials of construction, such as steel, wood, concrete, and paper. However, materials of construction based on high polymers require special techniques for characterization. Work by devoted physicists for the past 40 years has continuously demonstrated an acute awareness of this point. Such devotion is still required in order to determine end use performance because when plastics are stressed mechanically, exited electrically, optically or magnetically, or activated thermally, the molecules rearrange themselves at a rate dependent on time, temperature, and other environmental conditions as well as material history. In mechanical tests, such molecular rearrangements are described as viscoelastic phenomena. In electric measurements such molecular rearrangements are described as polarization phenomena. Other general phenomena such as microstructural changes, photodegradation, crosslinking, polarization, etc., result from optical and magnetic excitations. Unfortunately, many test technologists choose to overlook such phenomena in favor of obtaining oversimplified single point data.

Physical testing, in phenomenological terms, involves measuring the output response of a polymer system under given boundary conditions and in a given surrounding environment when it is subjected to external input physical excitations, as shown in Fig. 2. These excitations may be such everyday occurrences as force, displacement, heat, voltage, or luminous radiation. These may be in many simple wave shapes such as steps, sinusoids, ramps, pulses, gates, etc. or may be complex combinations of these simple forms. The amplitude may be small so that molecules are simply displaced, or it may be large enough to rupture the bonds between the molecules. Boundary conditions, which also influence the response, may be considered as geometric constraints placed on the specimen or any conditions that must be mathematically satisfied at the specimen faces. The time-form of the excitation functions, the boundary conditions, and the surrounding environment have enormous influence on the nature of the response, and therein lies the source of the apparent complexity and confusions associated with many of the test methods [*3*]. It is one of the main tasks of the physicist working with plastics to express this complex behavior in terms of a few identifiable, measurable, and useful parameters.

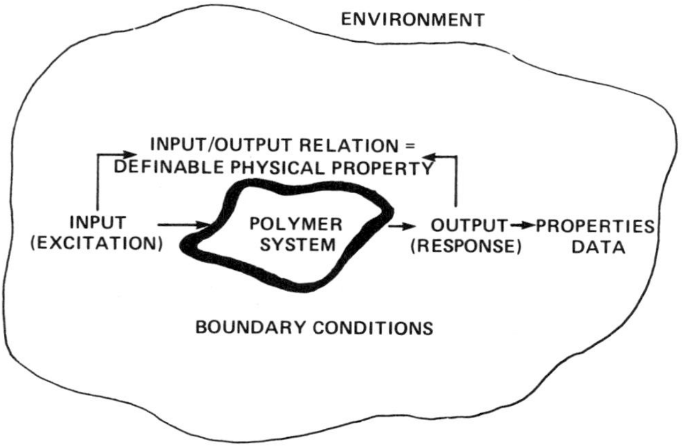

FIG. 2—*Physical testing can be represented as an input/output system with specified boundary conditions (after Turner* [3]).

There are two distinct approaches available for defining a physical property. One can select a particular type of excitation function and merely display the response function as a "physical property," or one can measure the input/output relationship. For example, consider the measurement of tensile strength where only the output response is measured, and modulus where the input and output are measured and the ratio is calculated. In any case, for a complete characterization of a physical attribute, the output response must either be independent of the boundary condition or must be determined under various boundary conditions. Also, the environmental conditions must be considered and the observed responses interrelated. For all physical tests, regardless of specificity, using the integrated or systems approach which is simply an input/output measurement under given boundary conditions and in a given environment, will remove the vagueness surrounding many tests. The general systems approach to physical testing requires the materials researcher to measure the output response for a material over a range of input excitations, environments, and boundary conditions.

The general representation of a physical property, in this case the dynamic mechanical damping of polymethyl methacrylate (PMMA), is shown as a function of both frequency (input excitation) and environment (temperature) in Fig. 3. In this case the test is independent of the boundary conditions. Another example of the general representation of a physical property is shown in Fig. 4. In this case, the Izod impact strengths of two thicknesses of polycarbonate are plotted against the notch radii, that is, the boundary conditions. Here the input excitation is fixed (and too limited for effective evaluation) because of machine design.

Practical data collection is usually abbreviated as necessary. Taken to the extreme, the above curves degenerate into single datum points.

Types of Test Methods

Test methods employed in physical testing can be divided into three classes: fundamental tests, component or practical tests, and hybrid tests.

Fundamental Tests

Fundamental tests yield data concerning basic physical properties. They permit reliable calculations under simple conditions that can be analyzed and expressed in absolute units. These tests provide a means for defining and evaluating the physical properties independent of the design and dimensions of either the test specimen or the test equipment, that is, the boundary conditions. However, fundamental tests are dependent on the environment and on the shape and size of the input excitation. Also, to be meaningful the processing conditions used to obtain the test specimens should be the same as the processing conditions that would be used to fabricate the article. An

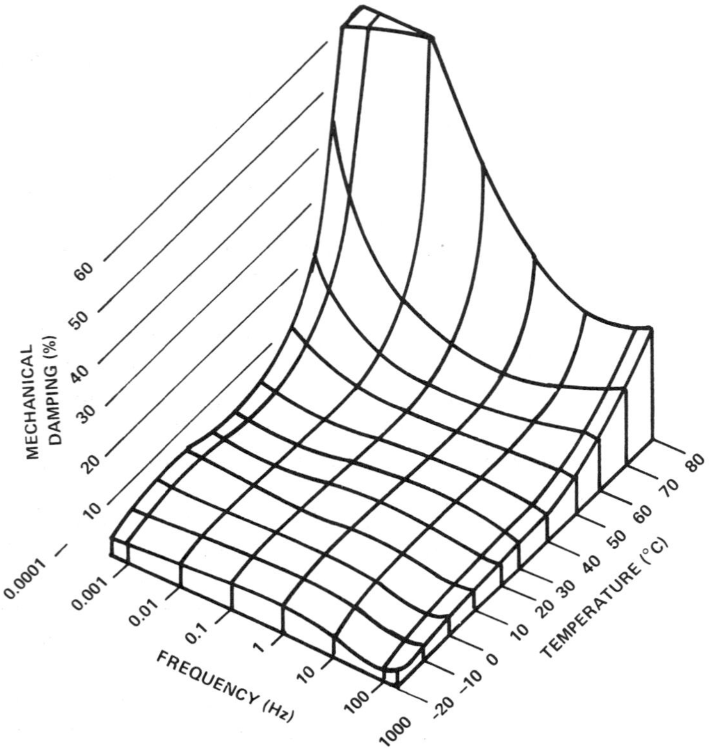

FIG. 3—*General representation of a physical property, in this case the dynamic mechanical damping of PMMA, as a function of both frequency (time) and temperature. Practical data collection is usually abbreviated as necessary. Taken to the extreme, the curve degenerates to a single datum point.*

alternative would be to obtain test specimens directly from the article. The loss tangent test data presented in Fig. 3 illustrate a fundamental type of test.

Component Tests

Component tests yield data from a finished part to allow the prediction of article behavior in service conditions. While this may be the most desirable way to test, in many cases component parts are usually not available, or the component cannot be tested *per se*. Also, component tests that provide simple "go–no go" data are usually slow, expensive, and so inaccurate that many are required to give reliable average results. The main restriction, however, is that component properties may be related to more than one fundamental physical property and be sensitive to processing conditions. When failure occurs it is not possible to determine which fundamental property or processing conditions must be improved. Component tests are vastly improved, however, when instrumentation is available for data acquisition so that combined effects may be isolated and specific properties calculated from the component data.

Practical Tests—These tests are special forms of component tests. In a practical test, a specimen is modeled so that the boundary conditions of the test are identical to those in which the component would be exposed. They are valid for specific applications only. While not as expensive as component tests, they are still limited because failure may depend on more than one fundamental property. Examples are bearing strength measurements, where load is applied to a panel by means of a pin having the same diameter and placement as the end-use application.

Hybrid Tests

Hybrid tests yield data from special specimen geometries that are used for obtaining general information supposedly simulating particular service conditions. The measured physical property is dependent on the design and dimensions of either or both the test specimen and test equipment. However,

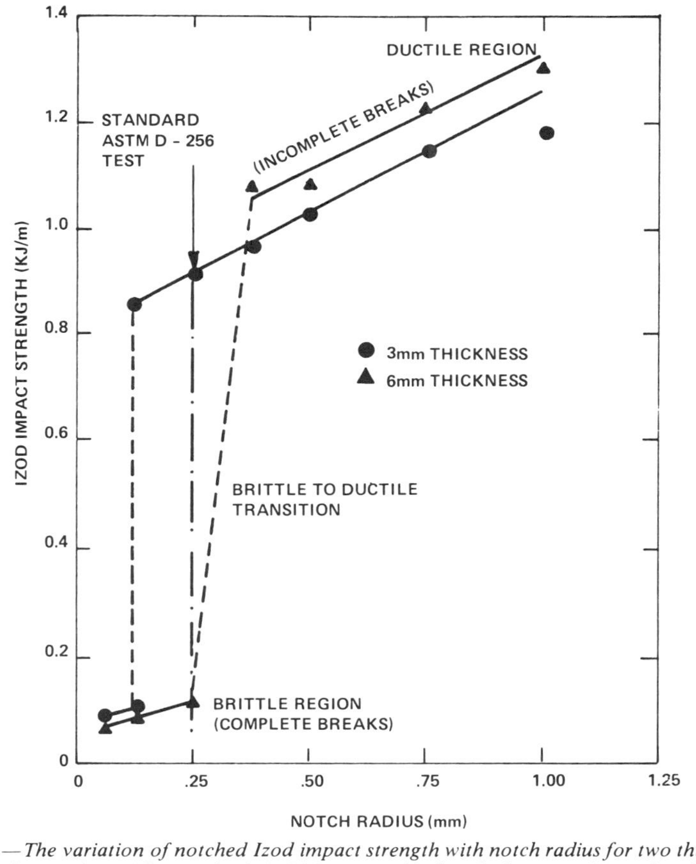

FIG. 4—*The variation of notched Izod impact strength with notch radius for two thicknesses of polycarbonate bars. The standard ASTM test would classify the 6-mm bar as brittle and the 3-mm material as ductile.*

these tests have no physical validity because the output response is dependent on the boundary conditions, and the boundary conditions imposed on the specimen are not the same as those imposed on the component during actual use. Examples are scratch resistance tests using scribers cut from arbitrary materials and of arbitrary geometry. The hybrid test can be considered as a practical test for only one specific set of boundary conditions. The Izod test data presented in Fig. 4 illustrate a hybrid type of test.

The selection of the type of test method used is governed by the reason for testing. For example, in quality assurance testing the measurement of variability is more important than the significance of the absolute result. Thus, hybrid tests are acceptable. Methods to obtain typical values can also use hybrid tests although fundamental tests are preferred. However, for research purposes and to establish suitability, fundamental or component tests are recommended whenever possible even if they are more time-consuming and more expensive than a hybrid test. A sound philosophy of engineering measurement is dependent on this.

Despite warnings, there still exists a great tendency among many to ascribe generality to hybrid tests for research and suitability testing. This practice is discouraged. The importance of avoiding the pitfalls associated with hybrid tests has been recognized by other experimentalists [*3-5*]. For example, the following is part of a 1937 commentary by Mooney [*5*].

> A hybrid test is one that stands somewhere between the practical and the scientific. It does not represent faithfully any particular service condition; and it is not designed so as to permit the calculation of any basic physical property. Most such tests in common use have been designed with half an eye on service conditions and an eye and a half of speed and accuracy of the laboratory test. If the measurements have any scientific value, it is largely a result of chance, or oversight.
>
> Concerning the hybrid type of test, there is little, in the author's opinion, that can be said to justify it. If the test neither resembles closely any service conditions nor yields data concerning the basic physical properties, the result will neither predict with certainty the behavior of the material in complex service conditions, nor permit reliable calculations in simple conditions that can be analyzed.

An example will serve to illustrate how each of these three methods could be selected by an experimentalist trying to solve the following problem: A lavatory sink made of calcium carbonate-filled polyester may crack during its service life because of repeated thermal shock when first hot and then cold tap water impinges on the bottom of the bowl. The best resin/filler system must be selected to obtain the maximum cycles to failure.

Examples of how the three classes of tests could be employed would then be:

Fundamental Test

Small test coupons could be tested for fracture toughness, modulus, thermal expansion, and specific heat and the results of these tests combined to calculate thermal shock resistance.

Component Test

The sink, or a model thereof, could be molded and subjected to thermal cycling by partially filling it alternately with water at the two temperature extremes until failure occurs [6].

Practical Test—The central area of a large test coupon containing a hole could be thermally shocked with water between temperature extremes until a crack initiates at the hole.

Hybrid Test

A small test coupon could be cycled by totally immersing it in beakers of hot and cold water and examining it periodically for cracks. Other tests could involve measurement of the Izod impact strength, etc.

Obviously, the boundary conditions of the hybrid test do not represent a real situation since no mechanical constraint is offered against expansion or contraction and the temperature distribution is uniform. Thus, any results obtained would have doubtful value. Component or practical tests represent tests with boundary conditions encountered in end-use, but several hundred cycles may be required to cause rupture. The fundamental test, on the other hand, permits the calculation of thermal shock resistance using small laboratory specimens without the need to construct a large component for each material composition considered. Using the fundamental type of test for at least initial screening also minimizes the overall spectrum of materials that can be considered as candidates for additional study as a finished part. This example also serves to illustrate an important principle: Fundamental tests can be developed to measure in absolute units any clearly defined physical property possessed by the test material under normal conditions. Such tests also will give more complete and more reliable information than hybrid or component tests.

Assessment of Standard Test Methods

The total number of organizations engaged in the development of standards in various subject areas exceeds 400 [7] in the United States alone. In particular, the American Society for Testing and Materials (ASTM) has issued approximately 4000 standard specifications and test methods. These methods cover a diverse range and include fundamental, component, practical, and hybrid methods needed for the various reasons for testing.

The number of these standard test methods is so large, and their status within national commerce is so long established, that their influence sometimes extends inadvertently beyond their proper domain. An important case has been the use of single-point hybrid test results in general design calculations and predictions. Such results were intended originally as quality control or practical values and certainly have no more extended applicability. An example of how a single datum point obtained in such a standard test can lead to an erroneous conclusion has already been illustrated in Fig. 4. When polycarbonate is notched and the impact strength measured, materials of both 3 and 6 mm thickness show a transition from brittle to ductile behavior dependent on notch radius. However, if only the standard ASTM Test for Impact Resistance of Plastics and Electrical Insulating Materials (D-256-78) was conducted on bars with a 0.25 mm radius notch, we would erroneously conclude that the material of 3 mm thickness is ductile while the material 6 mm thickness is brittle.

Many other standard tests have been developed to evaluate plastics in terms of fundamental properties. While most of these methods are scientifically sound, they still can be misleading. This is because standard methods by nature are arbitrary. Thus, the input excitation and boundary conditions of the test are controlled so that "materials properties" can be compared, or used as a basis for trade or acceptance. But these conditions are seldom relevant to the actual use conditions encountered. However, it is not the test methods that are to be criticized, but rather it is the failure by the experimentalist to extend these measurements over the correct time, temperature, or environment.

The development of standard design and performance criteria tests for plastics is in its infancy in ASTM Committee D-20. Tests methods such as the Standard ASTM Tests for Tensile, Compressive and Flexural Creep Rupture (D-2990-77) represent an excellent start in the proper direction. A greater awareness of the need for such tests and more input from designers will hopefully lead to greater activity in the future.

However, relevant standard tests cannot be expected to be available to cope with every situation. The researcher must learn to analyze the application and design meaningful physical tests to assess material performance. Hopefully, the discussion presented here will provide some new insights into how this can be done.

Selecting Relevant Physical Testing Procedures

Now that the basic rationale of physical testing has been reviewed, guidelines for selecting relevant physical testing procedures can be established. Such guidelines can be established for both the specific evaluation of a material for a known application and for the general evaluation of a material for potentially useful engineering attributes.

Evaluating a Material for a Known Application

This case is rather straightforward to outline although the actual test selection procedure may require much insight. An obvious rule for this case is that testing for properties that have no relation whatsoever to the performance properties of the finished article must be avoided. A rational selection of test procedures might be:

1. Analyze the application to determine the physical disturbances to which the material will be subjected.

2. Estimate the shape and magnitude of the expected input functions, the boundary conditions, and the environment to which the material will be exposed as well as the material's expected service life.

3. Select fundamental or practical standard tests or develop special tests to match the preceding requirements.

4. Evaluate the physical properties of the material using these tests at the application-focused conditions and service times.

5. Repeat the evaluations using exaggerated and suppressed inputs and environments to establish bounds for acceptable and unacceptable performance.

General Evaluation of a Material

This case can best be evaluated as follows:

1. Develop a working hypothesis of the various physical attributes believed to be possessed by the material.

2. Select test procedures and specimens that can be employed over a wide range of input excitations and boundary conditions, avoiding tests that merely provide a single datum point or that are poorly defined in real physical terms. Select fundamental or practical tests whenever possible and avoid hybrid tests.

3. Evaluate physical properties over a range of temperature and time. Establish bounds for acceptable and unacceptable performance.

4. Measure physical changes during and after isothermal or environmental aging. Establish bounds for acceptable and unacceptable performance.

5. Present data in a form useful to the designer.

Summary

A review of the methodology necessary for selecting physical testing procedures that provide data correlating with end-use performance has been presented. Selection of proper test methods requires that the shape and magnitude of the input functions, the boundary conditions, and the environment to which the material will be exposed match those of the end-use application. Hybrid types of tests that provide single point data are of no use in predicting end-use performance and should be avoided at all cost.

It is hoped that the discussion presented here has provided some guidelines that will be helpful in selecting test procedures that correlate with end-use performance. The reader may find it informative to analyze the tests discussed in the other papers presented in this Special Technical Publication to determine how the success or failure of these tests can be related to test selection rationale presented here.

References

[1] Von Krevelen, D. W., *Properties of Polymers, Their Estimation and Correlation with Chemical Structure*, Elsevier, Amsterdam, Oxford, New York, 1976.

[2] Beatty, J. R. and Sitz, C. E. in *Applied Polymer Science,* J. Craner and R. W. Tess, Eds., American Chemical Society, Washington, D.C., 1976, p. 100.
[3] Turner, S., *Mechanical Testing of Plastics,* CRC Press, Cleveland, Ohio, 1973.
[4] Alfrey, T., Jr., in *Mechanical Behavior of High Polymers,* Vol. VI, Appendix IV, Interscience, New York, 1948, p. 565.
[5] Mooney, M. in *Symposium on Consistency, ASTM STP 34,* American Society for Testing and Materials, 1938, pp. 9–14.
[6] Sherlock, J. M., *Plastics Engineering,* March 1976, p. 35.
[7] Brown, W. E., Frost, F. C., and Willard, P. E. in *Testing of Polymers,* Vol. I, J. V. Schmitz, Ed., Interscience, New York, 1965, p.1.

P. Flueler,[1] D. R. Roberts,[1] J. F. Mandell,[1] and F. J. McGarry[1]

Fracture Toughness Testing of HDPE Plastic Pipe

REFERENCE: Flueler, P., Roberts, D. R., Mandell, J. F., and McGarry, F. J., "**Fracture Toughness Testing of HDPE Plastic Pipe,**" *Physical Testing of Plastics—Correlation with End-Use Performance, ASTM STP 736,* R. E. Evans, Ed., American Society for Testing and Materials, 1981, pp. 15–31.

ABSTRACT: A program to investigate the applicability of fracture toughness test methods to high density polyethylene pipe is described. Fracture toughness test specimens cut from pipe material were tested in a precracked three-point bending test at different rates. The load versus crack opening displacement record was used in an attempt to judge the validity of the test and to determine the fracture load. This method appears to need further development for this class of materials, as the results were difficult to interpret. Use of the maximum force for tests which appeared to fail in a quasi-brittle manner resulted in calculation of a fracture toughness value which generally did not satisfy validity conditions. This value was used to predict the burst pressure of pipes containing surface cracks in the longitudinal direction. The prediction of burst pressure was slightly conservative, but in agreement with the trend of the experimental results. Fractographic analysis indicates three modes of crack propagation: ductile, quasi-brittle, and brittle.

KEY WORDS: plastic pipe, high density polyethylene, fracture toughness, fractographic analysis

Brittle fracture is a mode of failure observed under certain conditions in all classes of solid materials. Because of its catastrophic nature, it is usually the most undesirable failure mode. The technology associated with linear elastic fracture mechanics has been developed to such a sufficient degree in high-strength metals, that the fracture toughness can be used as a material property for design and material development purposes in many cases [*1,2*].[2] This allows the safe use of potentially brittle materials containing unavoidable flaws, and has led to more fracture resistant materials. The use of fracture mechanics also provides a context for the treatment of other, more stable, crack phenomena such as fatigue crack growth and stress corrosion cracking. Polymers, like metals, range in behavior from very brittle to very ductile, and the basic approach of linear elastic fracture mechanics has been applied to

[1]Head, Plastic Materials Department, Swiss Federal Laboratories for Testing Materials and Research, Dubendorf, Switzerland, graduate student, research associate, and professor, respectively, Massachusetts Institute of Technology, Cambridge, Mass. 02139.

[2]The italic numbers in brackets refer to the list of references appended to this paper.

some of the more brittle polymers [*3,4*]. Proper application of fracture mechanics to polymers could provide the basic toughness material property which is not generally obtainable from standard notched impact testing or instrumented impact tests if they do not incorporate a naturally sharp crack.

The use of linear elastic fracture mechanics requires 1. knowledge of the stress intensity factors, K, for the geometry of interest; 2. determination of the force at which unstable crack growth occurs; and 3. judgment as to the validity of the test result. The stress intensity factor for a particular geometry may be often found in handbooks on the subject [*5,6*], or may be determined by an appropriate stress analysis method. Determination of the fracture force and validity of the test can be complex, and methods suitable for metals may not be directly applicable to plastics. The validity question is critical, since simple calculation of the fracture toughness may give erroneous, nonconservative values if the fracture is not sufficiently brittle due to specimen size, test conditions, or precrack introduction procedure.

Once determined, the fracture toughness may be of direct use in materials development as a material property analogous to strength or modulus. If the fracture toughness is to be used in the design or analysis of a particular application, then it must be coupled with a determination of existing flaws in the actual item of interest. For example, if flaws larger than some size c_o can be detected by inspection, then the value of K for flaws of this size must be kept below the critical value, K_c (the fracture toughness) to avoid failure, with some margin of safety, under expected use conditions. Slow, stable growth of the existing flaws under load and environmental attack, which could cause them to exceed c_o, should be also considered.

This paper describes results of a study in progress to apply fracture mechanics to the failure of high-density polyethylene (HDPE) gas and water distribution pipes. Initial testing used small specimens similar to standard fracture toughness test specimens, but cut from the pipe wall, to determine the fracture characteristics, test validity, etc. These results were then used to predict the burst pressure of actual pipes containing known flaws. Cracks were generally oriented parallel to the extrusion, as this is normal to the maximum stress direction in pipes. Directional effects were determined on extruded flat sheets.

Experimental Methods

Test specimens were cut from extruded sheets or pipes in the desired geometry such as shown in Fig. 1. Cracks were introduced by first notching with a shaped saw blade, followed by razor blade sharpening, forcing the blade into the notch tip to the desired crack length using a universal testing machine. All fracture toughness specimens were loaded in three-point

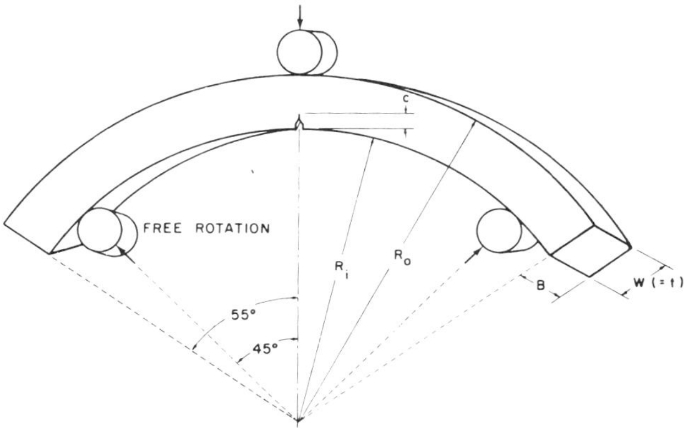

FIG. 1—*Specimen for 3-point bending of section of a cylinder.*

bending, as indicated in Fig. 1. Most of the specimens were cut from the circumferential direction of pipes as indicated, but some were cut from sheets and tested as notched rectangular beams. Specimens were loaded at various rates in either a universal or servohydraulic testing machine. The crack opening displacement (COD) was measured with a specially prepared gage similar to that described in Ref *7*, but with a greater displacement capacity. Unless otherwise noted, testing was done in an air-conditioned laboratory environment.

Full-scale pipe burst tests were done with as-extruded and precracked pipes. Razor-sharpened cracks of various length and depth were introduced on the outside surfaces of the pipes in the longitudinal direction. Pressurization was at a constant flow rate with water; the ends of the pipe were free to expand.

The materials used were HDPE in the form of 12.7-mm-thick extruded sheet, water pipe 200 mm outside diameter, and gas pipe 114 mm outside diameter. The water pipe containing carbonblack was used for the full-scale tests. The dimensions and properties are given in Table 1; it should be noted that the gas pipe is of lower density than the other two materials.

Stress intensity factors, K_I, were calculated from K–calibrations given in Ref *5* for bending specimens given in Fig. 1 and for pipes with cracks on the outside surface. The K–calibration for bending tests with the crack on the outside surface was obtained from P. G. Tracy of the Army Materials and Mechanics Research Center, Watertown, Mass. The calibration for the flat beam specimen used with extruded sheet material was taken from the ASTM Standard Test Method for Plane Strain Fracture Toughness of Metallic Materials (E 399-78*a*). The terminology used in this paper for the critical stress intensity factor is K_Q, where the subscript indicates that the validity of the test is not certain.

TABLE 1—*Materials and specimen geometrics*

Material (all HDPE) Origin	A Water Pipe	B Gas Pipe	C Extruded Sheet
Density (g/cm^3)	0.967	0.944	0.968
Carbon black (%)[b]	2.8	. . .	. . .
Yield stress, MPa	21.6	19.3	27.4/27.3[a]
R_o, mm	98.5	57.4	—
W, mm	18.0	10.8	25.4
c, mm	9.0	5.4	12.7
B, mm	18.0	10.8	12.7
Span, mm	. . .	. . .	102
$K_{I/P}$[c], $m^{-3/2}$	8000	10 400	5534

[a]Parallel/Perpendicular to extrusion direction.
[b]Material B has an unknown orange/pink colored pigment, Material C has no pigment.
[c]K-calibration, where P is the force applied at the center.

Results and Discussion

Crack Opening Displacement

The validity of plane-strain fracture toughness tests of metals is judged in part by the nonlinearity of the load versus the COD curve, which may result from excessive plasticity or stable crack growth (E 399-78*a*). The load-COD curve may have any of several basic shapes. Figure 2 is a schematic of the

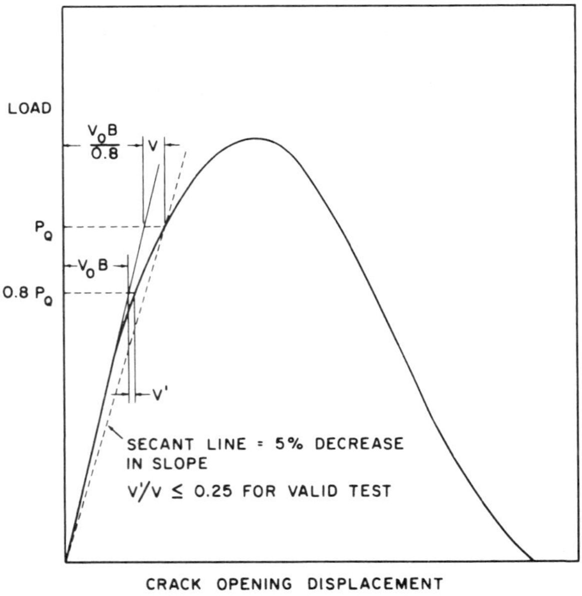

FIG. 2—*Schematic of COD linearity criterion for metals.*

shape commonly observed for HDPE, including the condition for a valid test. This condition was empirically derived for plane-strain tests on metals (E 399-78*a*), but its applicability to HDPE or other polymers has not been reported to the authors' knowledge. A second, related condition for a valid test is that the ratio of the maximum force in the test, P_{max}, to P_Q not exceed 1.10.

Figure 3 gives load-COD curves for HDPE taken from extruded sheets. The load, P_Q, to be used in calculating the fracture toughness, (K_{Ic} if the plane-strain test is valid) is determined from the intersection of the 5 percent offset secant line with the curve. The P_Q determined in this manner averaged 86 percent of the maximum load for this material, outside the P_{max}/P_Q requirement. The physical interpretation of the 5 percent offset was studied by loading the specimens to various points on the load-COD curve, unloading, staining with a dye penetrant, observing the amount of crack growth. Figure 4 indicates that the crack begins to grow on the interior of the specimen, close to the value of P_Q, and before the maximum load is reached. The effects of extrusion direction shown for the crack growth after P_Q are also reflected in a more rapid drop of the load-COD curve after the maximum load is reached with the crack propagating parallel to the extrusion direction.

The P_Q values determined from Fig. 3 result in a calculated fracture

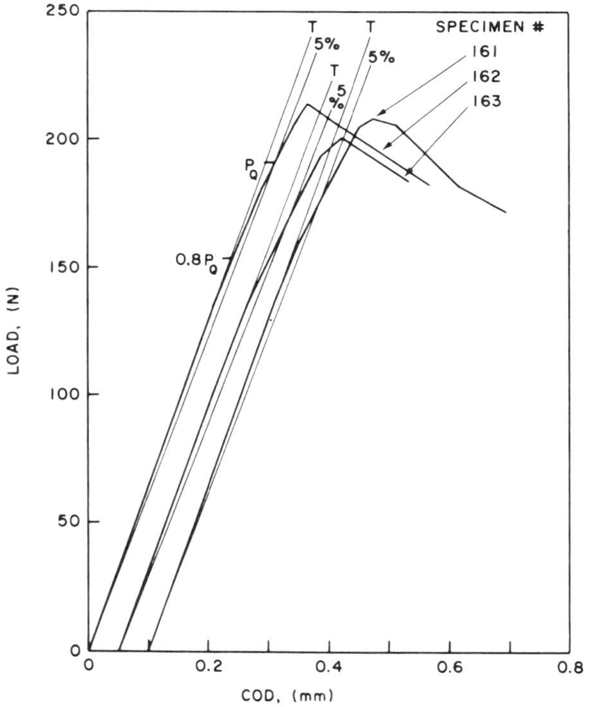

FIG. 3—*Load-COD curves and calculation of* P_Q *Material C.*

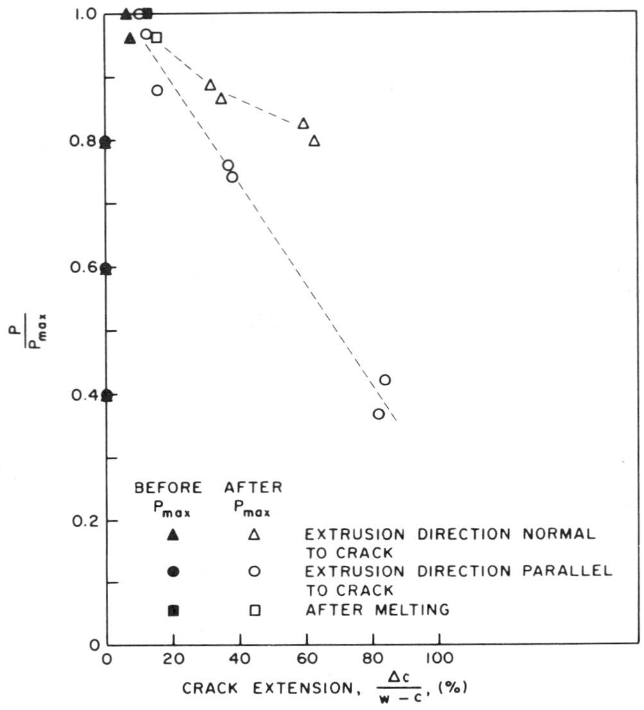

FIG. 4—*Load versus crack extension for Material C.*

toughness, K_Q, of 0.98 $MNm^{-3/2}$ (0.90 ksi $\sqrt{\text{in.}}$). A second validity condition is that a multiple of the calculated yield zone size be smaller than the thickness, crack length, and other dimensions (E 399-78*a*). This condition requires that

$$2.5 \left(\frac{K_Q}{\sigma_{ys}}\right)^2 < B, c, W\text{-}c \quad (1)$$

where σ_{ys} is the yield stress of the material. The left hand side of Eq 1, for the extruded sheet, averages 3.4 mm, well within the validity requirement. In addition to these validity conditions, the fracture surface was approximately 90 percent flat with some drawing of material near the surfaces. Taken together, these conditions indicate that the fracture toughness tests for the extruded HDPE sheet satisfy the validity requirements for plane strain fracture toughness testing of metals, with the exception of the P_{max}/P_Q ratio.

The results for the extruded sheet support the use of the metals technique for determining validity. However, tests of specimens cut from pipes of the two HDPE's given in Table 1, using the specimen shown in Fig. 1, gave contrary results. Although the mode of fracture and fracture surface appearance were similar to the sheet material if tested at high rates, the load-COD curves were very nonlinear as indicated in Fig. 5. Although the load-

COD linearity validity condition was satisfied, P_Q, determined by the 5 percent secant method, was only a small fraction of the maximum force. This clearly violates the P_{max}/P_Q limitation of 1.10, and Eq 1 is also violated if K_Q is calculated at P_{max}. Staining studies indicated that crack extension did not begin until loads much higher than P_Q. This problem was not related solely to the specimen geometry, because the same behavior was observed for standard specimens molded from these materials. These results indicate that the tests of Materials B and C do not give valid K_{Ic} values.

The load-COD curves for the sheet and pipe specimens indicate that problems exist in determining a valid K_{Ic} from HDPE fracture tests. Study of this problem continues, but the remainder of the fracture data in this paper were determined from the maximum load in the test.

Rate Effects

Determination of the effects of displacement rate on K_Q was difficult due to changes from obviously ductile, invalid tests, to brittle fractures in some cases as the rate increased. This trend was observed for both pipe materials. The extruded sheet material gave apparent brittle failures at all rates tested, and Fig. 6 indicates an increasing K_Q with rate in both directions. The qualitative results for the pipe materials suggest that at a transition to a higher K_Q, more ductile behavior is observed at sufficiently low rates.

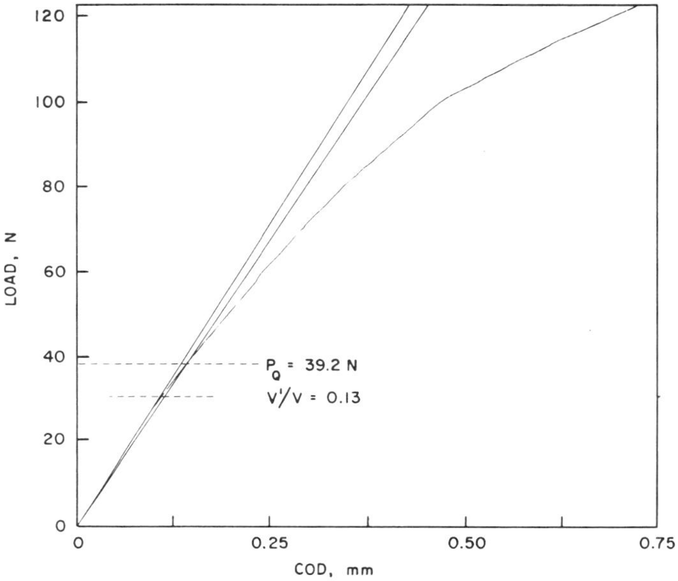

FIG. 5—*Load versus COD for Material B, loading rate = 6.7 MNm$^{-3/2}$/s.*

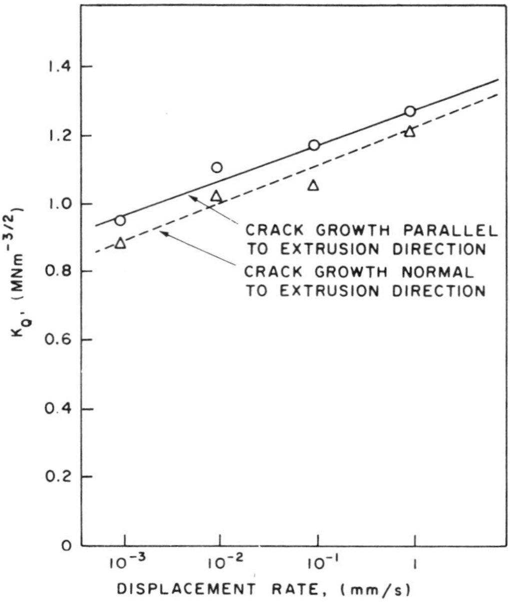

FIG. 6—*Rate of testing versus fracture toughness, Material C.*

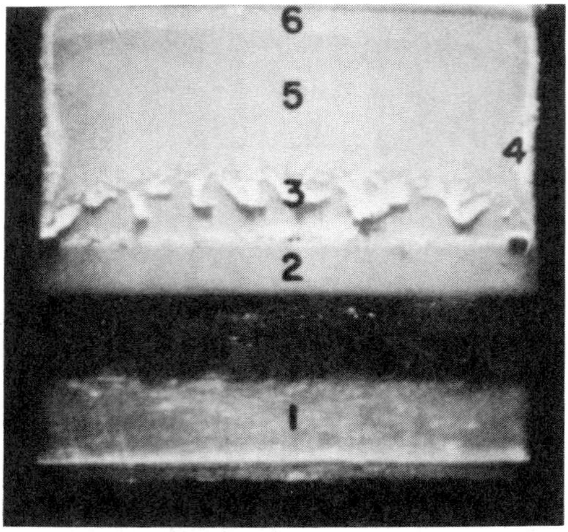

FIG. 7—*Fracture surface of Material B. Regions: (1) saw cut, (2) razor blade cut, (3) slow crack growth, (4) ductile edge zone, (5) quasi-brittle, fast crack growth, (6) crack deceleration near back surface.*

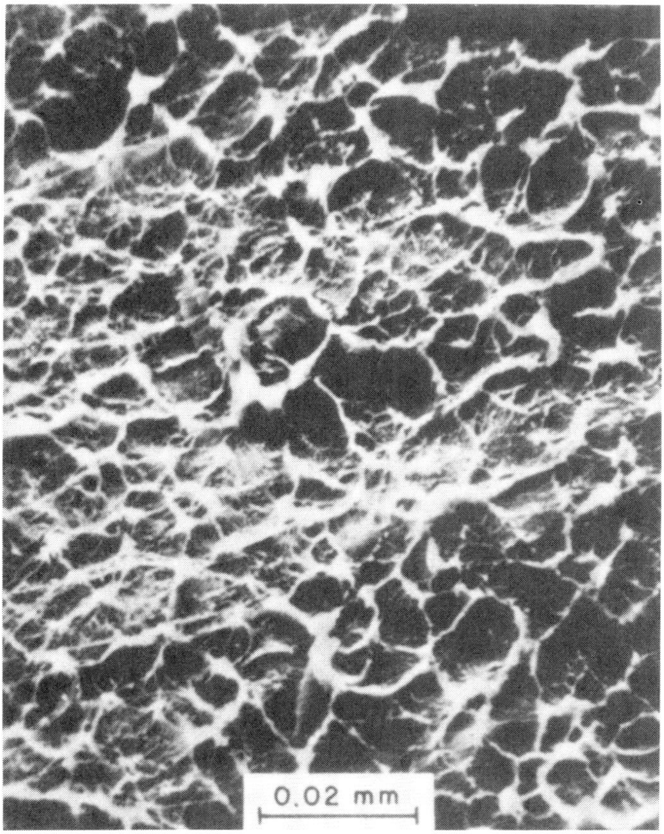

FIG. 8—*Quasi-brittle fracture surface of Material B (region 5 in Fig. 7).*

Mode of Fracture

Three distinct modes of fracture are observed under varying conditions:

1. Ductile: gross drawing and deformation of material.
2. Quasi-brittle: flat, brittle appearing fracture with ductility only on a micro-scale.
3. Brittle: flat, irregular features on fracture surface, no ductility observable up to 10 000 magnification.

Figure 7 shows a fracture surface from pipe Material B at low magnification. Most of this fracture surface was relatively flat, with ductile regions 3 and 4 at crack initiation, and near the edges. The interior flat region, 5, is shown in the scanning electron micrograph of Fig. 8. Here, the local ductility is evident, separating the surface into small regions on the order of 10 μm. This mode of growth is termed quasi-brittle. The features in Fig. 7 were also observed for

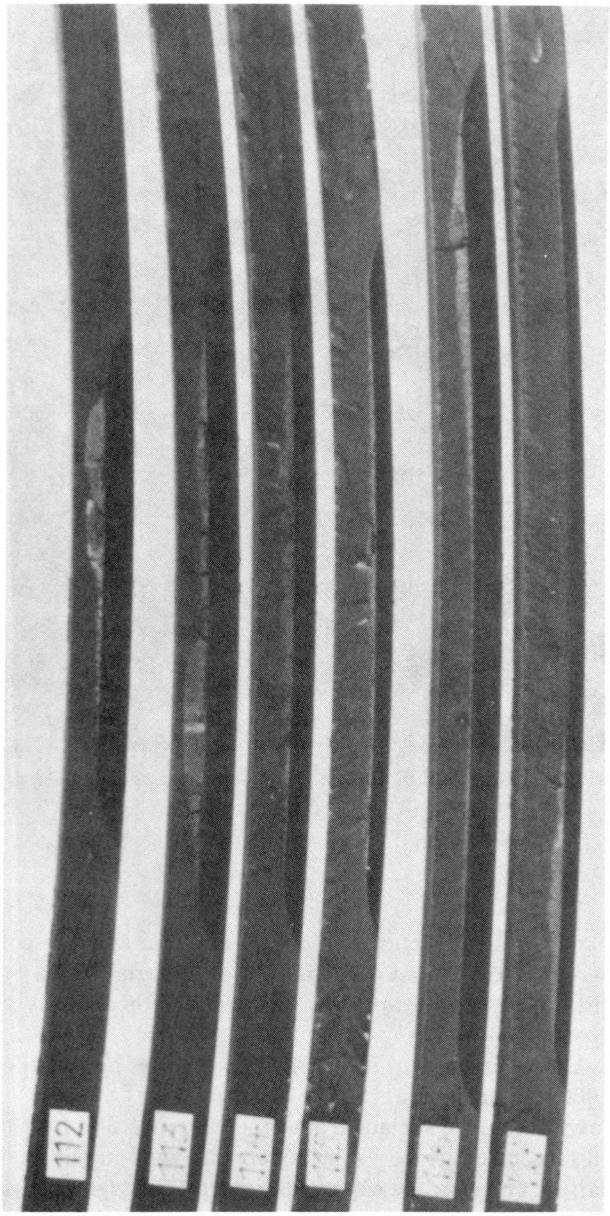

FIG. 9—*Fracture surfaces of precracked pipes, with precrack at bottom edge of each piece.*

the sheet material with the exception of the ductile region 3. The scale of the micro-ductile areas was similar to Fig. 8.

Figure 9 shows fracture surfaces from precracked pipes (Material A), in the region of the precrack. The dark, smooth areas are the machined cracks of various length and depth. The whitened regions are relatively slow crack growth areas along the machined crack border. The remainder of the fracture surfaces represent rapid crack growth (the long axis of the crack is along the length of the pipe; the slight curvature along the length is apparently due to residual stress). Figure 10 is a magnified view of the whitened area of Pipe 113. Figure 11 shows the border of the whitened area, and Figs. 12 and 13 are magnified views of each side of the border. The slow growth is similar to the flat area of the other materials in Fig. 8, but with slightly larger scale ductility. The fast growth area (Fig. 13) is distinctly brittle in appearance, with no apparent ductility. Although neither Material A or B showed the completely brittle mode in small specimen tests, the transition from quasi-brittle to brittle was observed with Material B when tested in liquid nitrogen (Fig. 14).

The general conditions for the three modes are consistent with the effects of stress state, rate, and temperature expected for many materials. The ductile mode is observed at higher temperatures, lower rates, and in plane stress regions near the surface. The quasi-brittle mode occurs under plane-strain conditions on the interior, and at lower temperatures and higher rates than the ductile mode. Cracks in the quasi-brittle mode shift into the completely brittle mode as the crack accelerates, particularly at low temperatures. The scale of the ductile features in the quasi-brittle mode is of the order of the expected

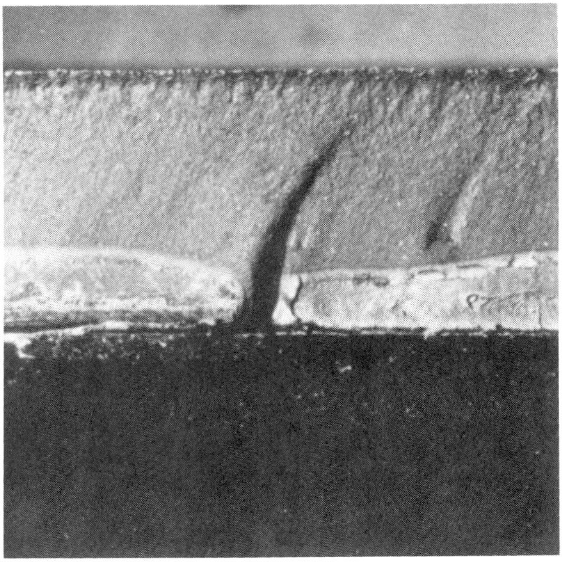

FIG. 10—*Magnified view of slow crack growth* (whitened) *region of Pipe 113.*

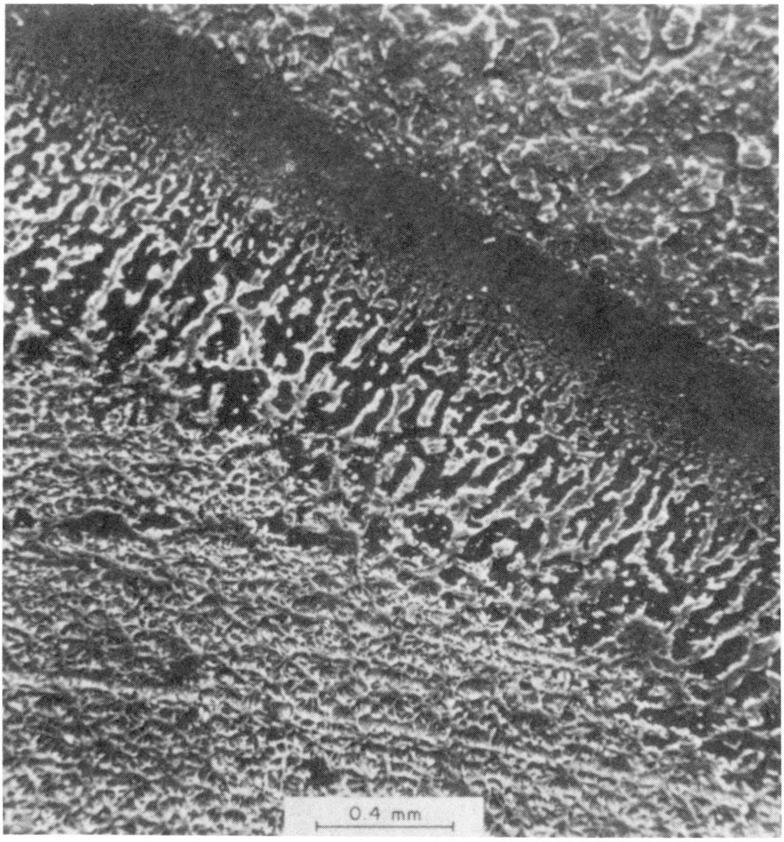

FIG. 11—*Border of slow* (lower left) *and fast crack growth.*

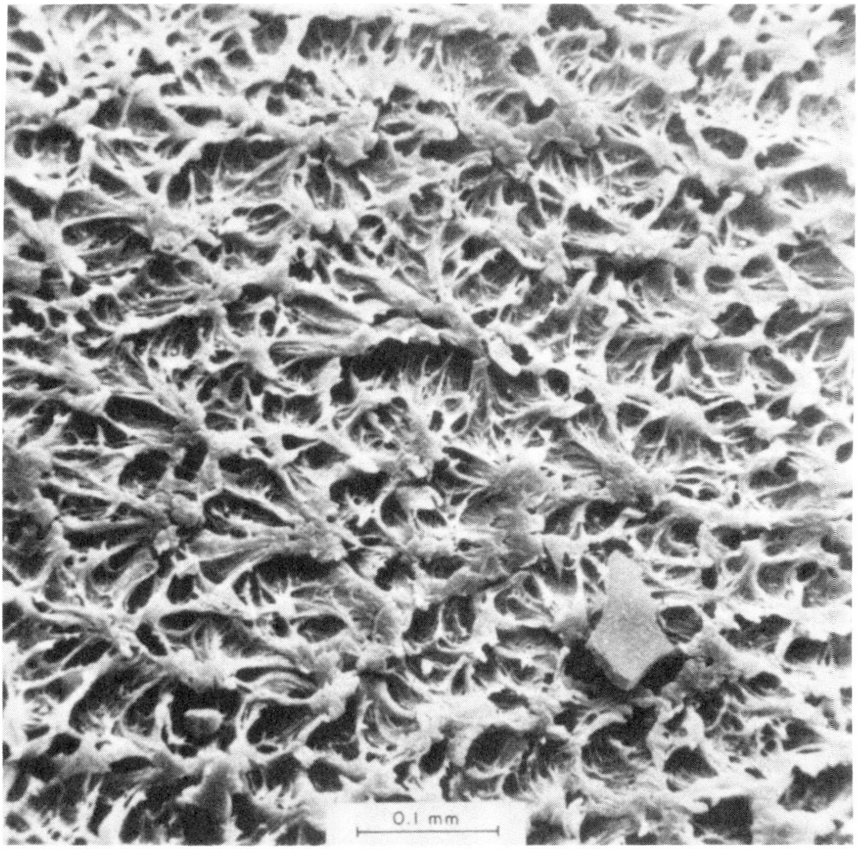

FIG. 12—*Slow growth region of Fig. 11.*

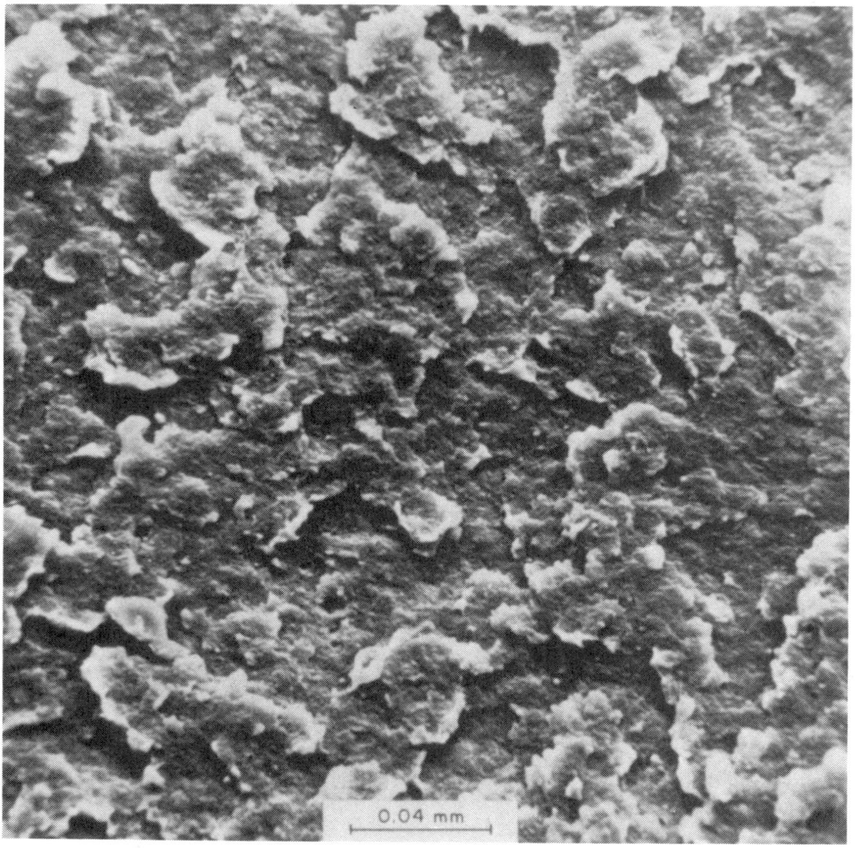

FIG. 13—*Fast growth region of Fig. 11.*

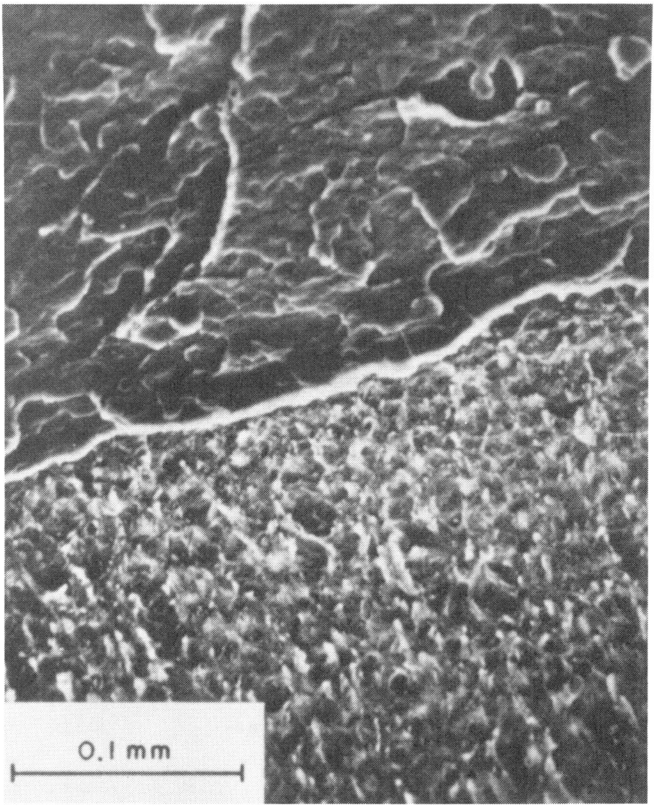

FIG. 14—*Slow* (bottom) *to fast crack growth border, Material B at* $-196°C$.

spherulite size. The brittle mode may represent interspherulitic growth as described for polypropylene in Ref *8*. Work is continuing on identification of the mechanisms of crack growth in each mode.

Precracked Pipe Burst Test Results

Figure 15 gives the results for the precracked pipe burst tests (Material A), and the predicted burst pressure based on small specimen tests. The value of K_Q from small specimens as shown in Fig. 1 was determined from the maximum load at a rate of 17 mm/s, at which the fracture surface was mostly flat, but other validity conditions were not satisfied, K_Q values for cracks on the inside and outside surfaces of small specimens were similar, and the overall average of 1.72 $MNm^{-3/2}$ was used in predicting the burst pressure of the pipes.

The results in Fig. 15 show approximate agreement between the fracture

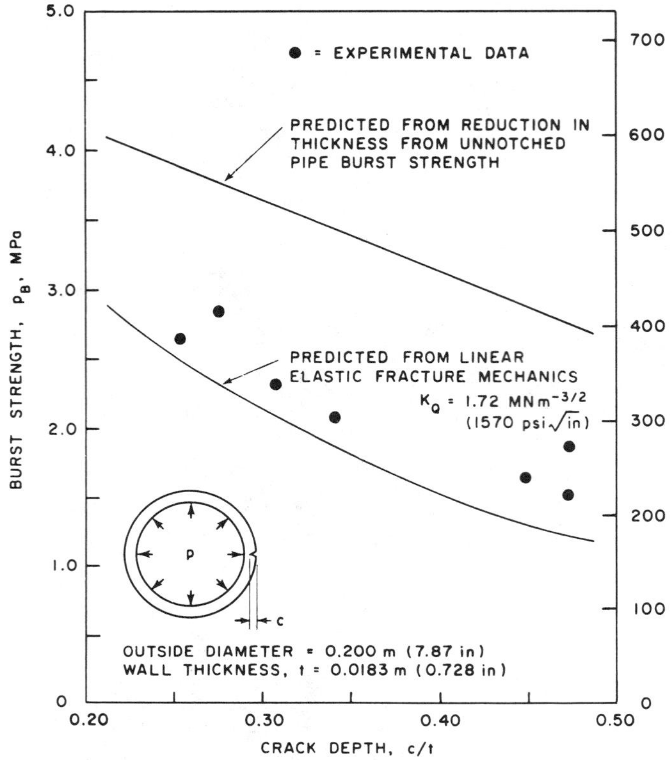

FIG. 15—*Predicted and experimental pipe burst data for HDPE pipe (Material A) with long longitudinal external cracks.*

mechanics prediction and the experimental burst pressure, with the prediction slightly conservative. A notch-insensitive prediction based on the reduction in wall thickness from the cracked and the uncracked burst pressure (ignoring bending effects) also shows the same trend as the experimental data, but is nonconservative. The cracks in the small specimens and in the pipes were too short to satisfy the conditions in Eq 1.

Conclusions

The validity of fracture toughness tests for HDPE pipe materials must be carefully considered; standard procedures used for metals may serve as a guideline. Modes of crack growth varied from ductile to quasi-brittle, to completely brittle, depending upon the degree of plane-strain constraint, the rate, and the temperature. Fracture toughness values determined from small specimens gave burst pressure predictions for precracked pipes which were in approximate agreement with experimental data, despite validity questions.

Acknowledgments

This work was sponsored in part by the Agency for International Development, United States Department of State. Mr. Flueler was supported by the Swiss Federal Laboratories for Testing Materials and Research while at the Massachusetts Institute of Technology.

References

[*1*] Knott, J. F., *Fundamentals of Fracture Mechanics,* Butterworths, Boston, 1973.
[*2*] Rolfe, S. J. and Barsom, J. M., *Fracture and Fatigue Control in Structures,* Prentice Hall, Englewood Cliffs, N.J., 1977.
[*3*] Berry, J. P., *Fracture, Vol. VII, Fracture of Nonmetals and Composites,* H. Liebowitz, Ed., Academic Press, New York, 1972, p. 37.
[*4*] Williams, J. G., *Polymer Engineering and Science,* Vol. 17, 1977, p. 144.
[*5*] Rooke, D. P., and Cartwright, D. J., *Stress Intensity Factors,* Her Majesty's Stationary Office, London, 1976.
[*6*] Tada, H., Paris, P. C., and Irwin, G. R., *The Stress Analysis of Cracks Handbook,* Del Research Corp., Hellertown, Pa., 1973.
[*7*] Friedrich, K., *Fracture 1977,* Vol. 3, Fourth International Conference on Fracture, Waterloo, Canada, 1977, p. 1119.

H. S. Loveless,[1] *L. R. Ingle,*[1] *and C. E. Gerrard*[1]

Fatigue and Its Relationship to End-Use Performance

REFERENCES: Loveless, H. S., Ingle, L. R., and Gerrard, C. E., "**Fatigue and Its Relationship to End-Use Performance**" *Physical Testing of Plastics—Correlation with End-Use Performance, ASTM STP 736,* R. E. Evans, Ed., American Society for Testing and Materials, 1981, pp. 32–44.

ABSTRACT: This paper describes two types of fatigue tests, cyclic and static, and their abilities to predict the long-term behavior of glass-fiber reinforced plastics in a marine and an electrical utility application.

Employing a modification of the existing ASTM Test for Flexural Fatigue of Plastics by Constant-Amplitude-of-Force (D671-78), a procedure was devised for differentiating styrene-suppressed polyester orthophthalic resin systems used as secondary bonding media for the encapsulation of wooden stringers to the hull section of marine craft. Using specimens machined from overlapped panels processed with various resin systems, the laboratory test employs a fixed stress at the overlap of 13.8 MPa (2000 psi). The resistance of the overlap to the completely reversed bending load is measured in terms of the number of cycles endured to failure by the overlap. Correlation of these laboratory measurements with actual service tests of pleasure craft in the Gulf of Mexico has been excellent, thus attesting to the validity and usefulness of such data for predicting the fatigue resistance of the bond between the stringers and hull section of the boat under a fatigue environment imposed by choppy waters in end use.

The paper concludes with a discussion of the Prot method for the prediction of the long-term tensile strength of plastics under static (fatigue) loading, together with an analysis of an electrical cable application of reinforced plastics, based upon the Prot test and elastic modulus considerations.

KEY WORDS: orthophthalic, reinforced plastics, composite, styrene-suppressed resins, styrene emission, encapsulation, hull stringers, chopped-strand mat, secondary bonding, dynamic, overlapped, ACSR cable, power transmission, aluminum conductors, core, steel-reinforced, cyclic fatigue, static fatigue, long-term strength, elastic modulus, RP pultruded rod, tower-to-tower, drop-cable, Prot tests, endurance limit, Wohler method, creep-rupture, load-bearing, laminate, end use, S-N curve, specimen, testing, polyester, chemistry

Two types of "fatigue" are generally encountered in service when structural materials such as reinforced plastics (RP) are used. These are dynamic (or cyclic) fatigue resulting from loads which induce vibration in the supporting member, and static fatigue, from the application of fixed static loads which must be sustained for very long times, often measured in years. This paper

[1]Research associate, PPG Industries Inc., Fiber Glass Research Center, Pittsburgh, Pa. 15230, materials research manager, Glastron Boat Company, Austin, Tex. 78766, and senior process engineer, Hercules Inc., Aerospace Division, De Soto, Kans. 66018, respectively.

describes two applications involving these respective types of fatigue, and the laboratory methods used to predict the performance of the particular RP composites employed under the actual end-use conditions.

Marine Application of Reinforced Plastics

The need to limit styrene emissions associated with unsaturated polyester resins in open contact molding has long been recognized by the marine industry as a safeguard to the health, morale, and safety of its employees. However, in recent years, pressure exerted by such agencies as the Occupational Safety and Health Association (OSHA) and National Institute for Occupational Safety and Health, (NIOSH) to reduce styrene emissions to levels of 100 ppm, or less, has led proponents of the open mold industry to further reduce levels previously tolerated by them in their production of large plastic components. This has been effectively accomplished by the Glastron Boat Company, for example, by controlling the release of styrene from the laminate in fabrication of pleasure craft such as the Model SSV-176 (Fig. 1). This was accomplished by the judicious employment of suppressed orthophthalic, general purpose resins. The use of a suppressant has been found to reduce the styrene weight loss in such resins by as much as 80 percent, depending upon the particular system used. The work described here concerns

FIG. 1—*The Glastron Model SSV-176 Runabout.*

TABLE 1—*Description of suppressed resins used in boat construction.*

Resin	Description	Styrene Emission, % by weight[a]
A	control or nonsuppressed	2.5 to 2.7
B	initial styrene suppressed	1.8 to 2.0
C	improved styrene suppressed	0.5 to 0.7

[a]Following a 2-h cure of 250-g resin casting under carefully controlled laboratory conditions.

the evaluation of (1) standard (that is, nonstyrene-suppressed), (2) initial styrene-suppressed, and (3) improved styrene-suppressed orthophthalic polyester resins, used as secondary bonding media in the mat strip encapsulation of wooden stringers to the hull of the boats. A description of these resins and their respective styrene emissions, determined as percent weight loss following a 2-h cure, can be found in Table 1. The chemistry of these suppressant systems is, of course, proprietary and cannot be revealed. The resins are, however, commercially available from various manufacturers.

The hull stringers, fabricated of pine, are shown in Fig. 2 prior to encapsulation. Encapsulation of the stringers to the hull of the boat is accomplished with chopped-stand mat strips saturated with the resin of interest during the process (Fig. 3). The secondary bonding characteristics of the resin are important here, since the hull surface to which the stringers are bonded is completely cured. The plywood flooring and the bulkhead of the boat are also mat stripped to the hull laminate in similar fashion.

In an effort to predict the secondary bonding characteristics of the various

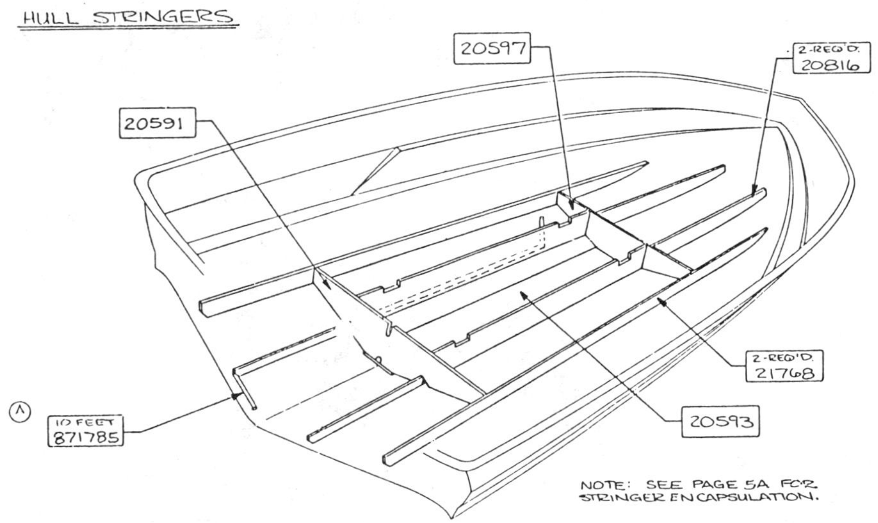

FIG. 2—*Skeletal view of the SSV-176 showing the hull stringers prior to their encapsulation.*

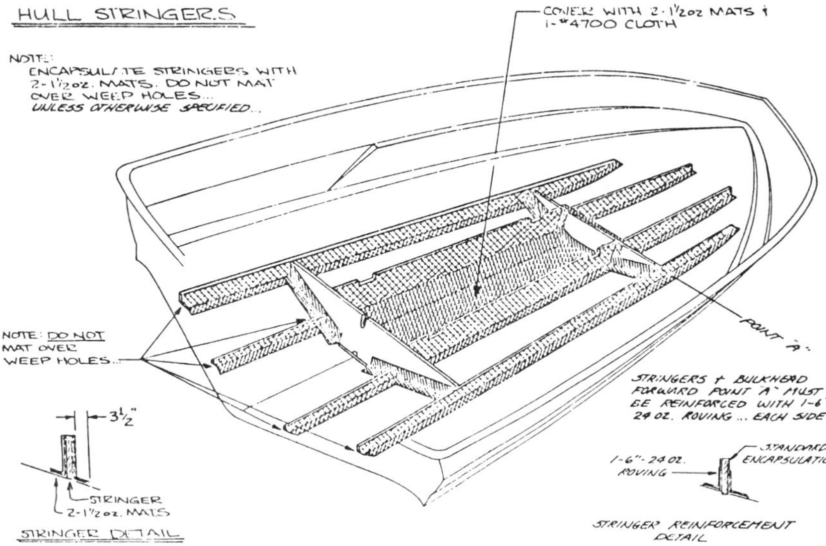

FIG. 3—*Skeletal view of the SSV-176 showing the hull stringers after their encapsulation.*

resins employed in the fabrication of boats, Glastron used a number of conventional static tests which proved inconclusive, and often meaningless, in terms of actual experience gained with test boats on the Gulf of Mexico. Among these were adaptations of the ASTM standards for flexural, tensile, and lap shear strengths. Since there was frequently a distinct lack of correlation between the information gleaned from such tests, and the damages incurred with test boats under service conditions, Glastron solicited the help of the Testing and Evaluation Laboratory of Pittsburgh Plate Glass (PPG) Industries toward developing a test that would more reliably predict secondary bonding problems sometimes encountered in end-use. These problems were mainly manifested by delamination of the mat strip encapsulation of wooden stringers, or braces, to the hull of the boat, thereby permitting flexing of the stringers to such an extent, that in some cases, hull punctures eventually developed. Since the stringers lend reinforcement and stiffness to the hull, in addition to providing a foundation for the plywood flooring of the boat, their role as structural members is at once obvious.

Subsequent discussions with Glastron personnel revealed that in various test boat hulls, the stresses that lead to hull failure were found to arise from cyclically loading and unloading the hull as it moved through rough or choppy waters, and from impacts of sudden hull reentry into the water after encountering a particular disturbance such as a wave. Hence, it was clear that the loads encountered in service were of a "dynamic," rather than "static," nature, and the most reasonable approach to procuring laboratory data showing some positive correlation with test boat observations lay in fatigue. Accordingly, suitable panels were prepared for test and forwarded to the Testing and Evaluation Laboratory, since Glastron lacked the fatigue equipment needed for this undertaking. To simulate actual stringer-to-hull bonding with the impregnated chopped strand mat strips, identical procedures were used in the preparation of the overlapped panels for test purposes. That is, overlap of the panels was achieved after allowing the first panel to cure for 24 h. Three sets of panels, all prepared with the resin systems of Table 1, were submitted for testing on three different occasions to establish validity and repeatability of the test results.

The fatigue test devised for this work by the Testing and Evaluation Laboratory entails a modification of the existing ASTM Method Test for Flexural Fatigue of Plastics by Constant-Amplitude-of-Force (D 671-71). Specimens used are of Type B geometry, and are machined from the laminates such that the overlap falls midway along the specimen's reduced section. The overlap is 20 mm (0.75 in.) long and is approximately 6.4 mm (0.25 in.) thick. Since the overlap represents twice the thickness of either of the two identical laminates used to prepare the test panels, it experiences only one-fourth the stress of the nonoverlapped sections within the reduced portion of the specimen for a given cantilever load. Selection of a stress level high enough to cause the overlap to fail, and yet low enough to enable the thinner sections of the specimen to survive under the imposed loading conditions was obviously a matter of immediate concern at the onset of testing. A trial-and-error procedure, augmented with some previous knowledge of the fatigue resistance of polyester chopped strand mat laminates, reduced the process of choosing the proper stress to a minimum. Unlike the ASTM standard, the test employed here involved a single stress level, instead of the normal complement of four or five. Hence the development of an S-N curve, as such, was never intended. The test simply employs a fixed stress at the overlap of 13.8 MPa (2000 psi). The resistance of the overlap to the completely reversed bending load is measured in terms of the number of cycles endured to failure by the overlap. Figure 4 shows the test arrangement, and Fig. 5 shows two randomly selected specimens before and after test. Some typical test results, based upon a minimum of 20 determinations in each case, are presented in Table 2.

As mentioned previously, overlapped laminates processed with both nonstyrene-suppressed and suppressed resins have been evaluated on three separate occasions over a one year period, and the results of these tests have consistently ranked these resins in the same order with respect to their secondary bonding characteristics. More importantly, the improved styrene-suppressed Resin System C of Table 1 has shown marked superiority over either of the other two Systems (A and B), both in terms of the laboratory fatigue test, and boat testing on the Gulf of Mexico for a total of 40 h in light, moderate, and heavy chop waters at slow, moderate, and fast speeds. The results of these field tests are summarized in Table 3.

FIG. 4—*Flexural fatigue test of overlapped chopped strand mat panel showing specimen tested as a cantilever beam under constant amplitude of force.*

Electrical Cable Application of Reinforced Plastics

The effect of fatigue, most frequently referring to the decline in strength of a material as the result of repeated stressing, is an important factor in modern engineering, since many machine components function under conditions of reversing and repeating loads. Long-term, static-load tests may be also

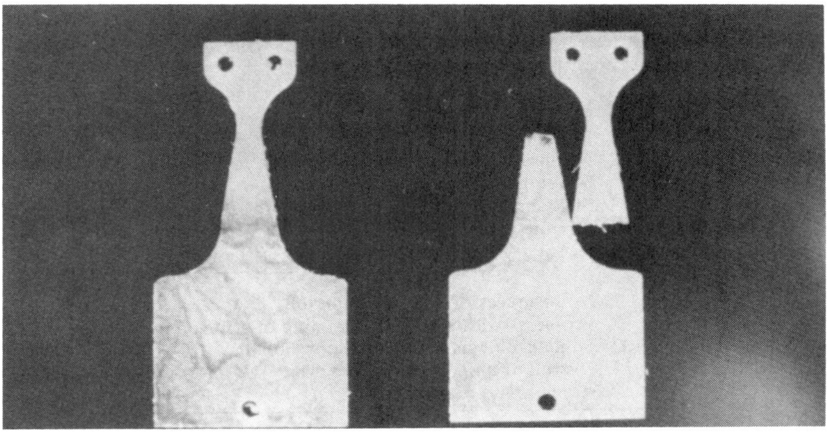

FIG. 5—*Failure of the overlap is characterized by complete debonding at the interface of the overlapped panels following the fatigue test.*

TABLE 2—*Fatigue resistance of overlapped mat laminates tested at 13.8 MPa (2000 psi) stress.*

Laminate Resin	Geometric Mean of Cycles for Overlap Failure[a]
A	259 000
B	103 000
C	1 092 000

[a]Tested as a horizontal cantilever beam at a frequency of 30 Hz under reversed bending loads in an environment carefully controlled at 22.8°C (73°F) and 50 percent RH. The results shown are for a minimum of 20 tests in each case.

regarded as a special case of fatigue, since "fatigue" occurs as the result of long-time loading. One of the best examples of static fatigue can be found in the steel supporting core of an aluminum conductor, steel-reinforced (ACSR) cable used in electrical power transmission. Cables of this nature are a familiar sight, stretching across open sections of the country on steel towers such as those shown in Fig. 6. A mathematical analysis of the catenary loading of such a cable under its own weight and other loads superimposed upon it by high winds and snow and ice, for example, shows that the steel core of the cable may sustain as much as three-fourths of the total stress borne by the cable. The long-term strength of the cable is thus critically dependent upon that of the supporting core, since most of the stress is carried by this component, by virtue of its much higher stiffness and strength, compared to the aluminum conductors. ACSR cable is used in many sizes, the larger ones such as code name "Thrasher," are comprised of as many as 76 aluminum conductors and 19 steel-reinforcement rods as the core (see Figs. 7 and 8). However, for

TABLE 3—*Comparison of resins with field data obtained from Gulf of Mexico test boat trials.*

Resin Used	Observations
A	1. No instances of delamination in hull. 2. No instances of stringer bracer torn loose due to poor secondary bonding. 3. Slight crazing problem on hulls due to severe flexing during test run. 4. Overall testing showed satisfactory performance in the area of secondary bonding characteristics of the resin.
B	1. No instances of hull delamination. 2. Problems with secondary bonding of stringers to hull. 3. Some problems with crazing on hulls. 4. Overall testing showed somewhat poorer results, particularly in strength and secondary bonding.
C	1. Superior hull laminate strength. 2. No problems with secondary bonding. 3. No crazing on hulls. 4. Excellent overall performance.

FIG. 6—*Electrical power transmission via high voltage ACSR cables.*

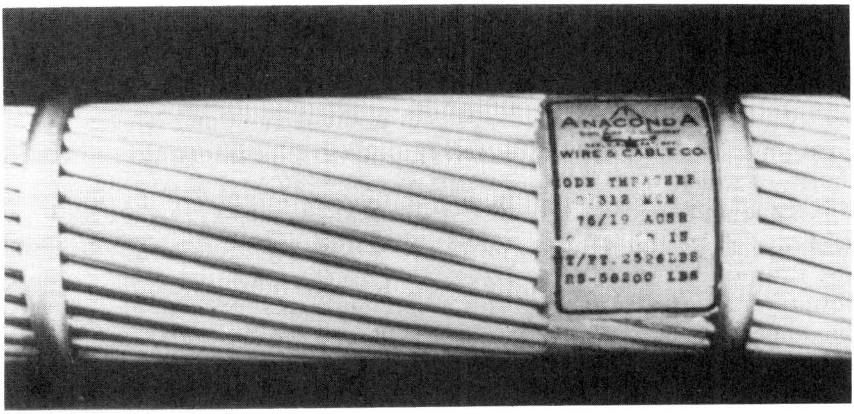

FIG. 7—*View of a 150 mm (6 in.) length of code "Thrasher" ACSR cable comprising 76 aluminum conductors and 19 steel-reinforcement rods as the core.*

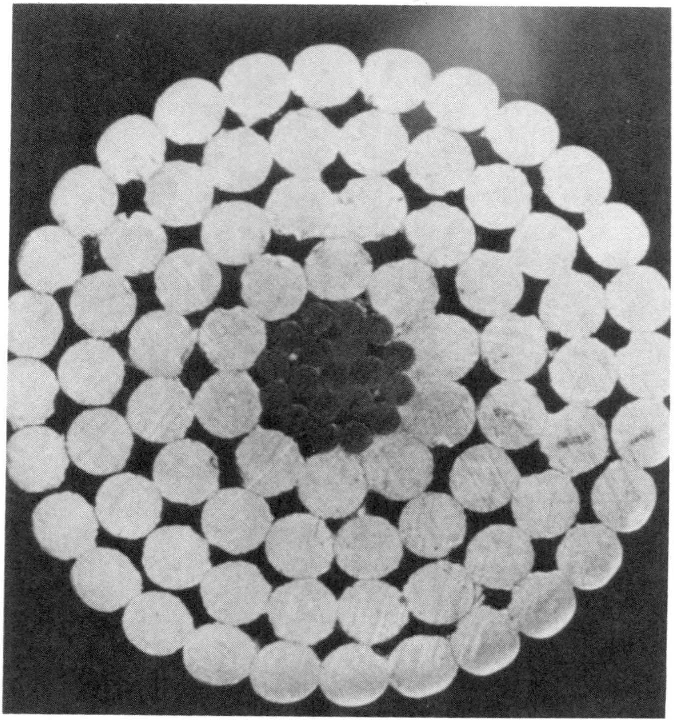

FIG. 8—*End view of the ACSR cable of Fig. 7 showing the cross sections of the conductor and core materials.*

purposes of this discussion, the balance of this paper will reflect treatment of the code name "Raven" cable, consisting of six aluminum strands and one steel, and the consideration given the possible replacement of the conventional steel core with a pultruded RP rod of the same 3.371 mm (0.1327 in.) nominal diameter. Interest in the replacement of the steel core of the ACSR cable with a plastic rod was expressed by the Wire and Cable Division of the Anaconda Company about five years ago, and stemmed from impending steel shortages which would seriously curtail the production of ACSR cable by Anaconda and similar industries. Smaller cables, such as the "Raven," are used for voltage applications in the 15 to 35 kV range in both tower-to-tower and drop-cable installations. Drop-cable usage entails the distribution of power from a primary voltage distribution source to a private residence over a maximum span of 60 m (200 ft), as opposed to tower-to-tower installations at spans of perhaps 360 m (1200 ft).

In order to establish the dependency and interaction of the respective cable components in an actual tower-to-tower installation, initial work on this project was concerned with catenary studies relative to code name "Drake" cable, consisting of 26 aluminum and 7 steel strands. An analysis of the mechanical properties of aluminum, steel, and the proposed pultruded RP rod, as fully expected, revealed substantial differences among them. The dependency of the cable upon the high elastic modulus of the steel core was particularly evident from the catenary analysis, and suggested that a problem would exist if a lower modulus core were substituted for the seven steel strands (or rods) comprising the core of the "Drake" cable. The transfer of tensile loads, for example, is critically dependent upon the tensile modulus of the core, which must assimilate these loads ahead of the aluminum conductors if the cable is to survive. Since the ratio of the respective moduli for steel and aluminum is approximately three, the steel core proves effective in supporting the weight of the cable suspended as a catenary in actual service. However, the roles of the core and conductor are reversed when a fiberglass rod is substituted for the steel core in an ACSR cable. This circumstance, again, arises from the difference between the respective tensile moduli of the two cores, but, in this case, the modulus of the core is only about one half that of the aluminum conductors. Hence, under tensile loading, the aluminum strands pick up the load in advance of the plastic core, thus negating the role of the core as the principal load-bearing component of the cable.

In an application such as this, it can be scarcely disputed that the long-term tensile strength of the cable's core is another property of prime importance, since ACSR cables have a normal life expectancy of 35 to 40 years. With this in mind, the respective long-term tensile strengths of the standard steel and the proposed reinforced plastic cores, used to strand the "Raven" cable, were determined by a technique known as the Prot method, described in a previous publication by Loveless and co-workers.[2] Unlike the conventional "Wohler" method, which consists of subjecting a number of specimens to different levels of stress to determine the length of time that a given stress can be sustained, the Prot test employs progressively increasing loads to achieve failure of the test specimens. The instrumentation used to provide such loads is shown in Fig. 9 and consists essentially of (1) a steel-shot metering device to provide the progressively increasing loads, (2) a reservoir for holding the steel shot in reserve, (3) an electrical timer to provide an accurate record of the duration of the test, and (4) a creep-rupture machine. A top view of the steel-shot metering device is shown in Fig. 10.

Estimates of long-term strength are obtained by testing a total of ten specimens of the sample of interest, employing two specimens at each of five different loading rates. An analysis of the failing loads and their respective rates of application provides a means of estimating the load, or stress, that can be sustained by the material indefinitely, that is, the estimated endurance limit. The validity of this approach has been established previously by a comparison of Prot estimates with literature values for long-time strength, as

[2] Loveless, H. S., Deeley, C. W., and Swanson, D. L., *Transactions*, Society of Plastics Engineers, Vol. 2, No. 2, April 1962.

FIG. 9—*General view of the Prot equipment used to determine the long-term tensile strength of the steel and RP cores of the code "Raven" ACSR cable.*

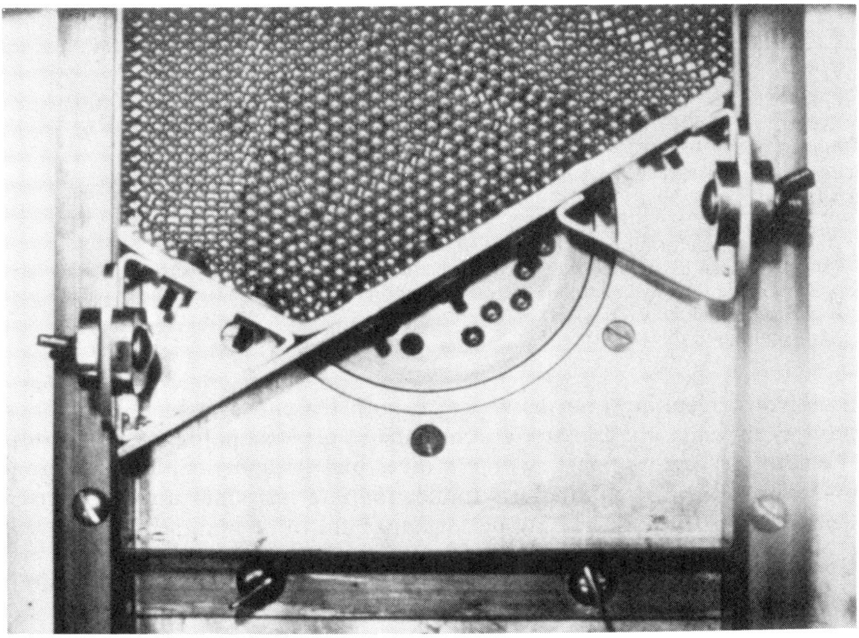

FIG. 10—*Top view of the steel-shot metering disc of the Prot apparatus used to provide progressively increasing loads to the specimen until rupture occurs.*

reflected by Wohler-type tests on a number of different materials.[3] The endurance limit, E, is found from the following equation by plotting αt_B^2 versus t_B, where α is the time rate of increase of stress, t_B is the time to break, S_o is the preload, and K is a constant. Such a plot yields a straight line, the slope of which is $2(E - S_o)$. Since S_o is known, E is readily found.

$$\alpha t_B^2/2 + (S_o - E)t_B = K$$

The results of the Prot tests on the respective "Raven" cores are given in Table 4, together with strength retention values calculated on the basis of their respective short and long-term strengths. It should be remembered that these tests were conducted under controlled laboratory conditions, and that outdoor weathering would have an additional influence on the core's ability to sustain long-term static loads. However, on the basis of the laboratory measurements described here, it can only be concluded that the steel core presently used in ACSR cables is a much better choice for this application than is the RP pultruded rod. While reinforced plastics have been, and will continue to be, used in structural applications more amenable to them than to metals for a myriad of reasons, misapplication of these materials could prove

[3]Loveless, H. S. and McWilliams, D. E., *Polymer Engineering and Science*, Vol. 10, No. 3, May 1970, pp. 139–147.

TABLE 4—*Long-term tensile strengths of steel and RP cores used in code "Raven" ACSR cable.*

Core Material	Tensile Strength				Strength Retention, %	Glass Content, % by weight
	Short-term[a]		Long-Term[b]			
	MPa	ksi	MPa	ksi		
RP rod	1300	189	720	104	55.5	73.3
Steel rod	1600	231	1450	211	91.3	na[c]

[a]Instron test result (test is concluded in about 1.5 m).

[b]Stress that can be endured (the so-called "endurance limit") for years without failure under long-term static loading conditions (as determined in an environment carefully controlled at 22.8°C (73°F) and 50 percent RH).

[c]na = not applicable.

disasterous, resulting in a disservice to both the end user and the plastics industry in which these materials continue to grow at unprecedented rates. The need for long-term testing of reinforced plastics under conditions closely approximating those of end-use application is clearly indicated, if reliable measures of their load-bearing performance are to be obtained.

Conclusion

In keeping with the theme of this technical seminar, "Determining the Suitability of Plastics for a Purpose," this paper has cited two applications of reinforced plastics which have entailed two different aspects of fatigue commonly encountered in end use. While the marine application of these materials is clearly vindicated by both the laboratory and field studies, the electrical cable application has been discouraged on the basis of laboratory data depicting the stiffness and strength retention of the proposed RP core.

Acknowledgments

The contributions of C. E. Daggerhart and J. S. Lail of PPGs Fiber Glass Division in conducting the tests described in this paper are kindly acknowledged. Special thanks are also due Andrew Wright, Senior Overhead Engineer, of the Wire and Cable Division of the Anaconda Company for many helpful discussions bearing upon the proposed ACSR cable application described in the paper. The assistance of Shelia Pendleton in typing the manuscript is also appreciated. Finally, the cooperation extended by both the Glastron Boat Company and the Anaconda Company during the preparation of the paper, and the permission of the Fiber Glass Division of PPG Industries, Inc. to publish it, are hereby gratefully acknowledged.

D. R. Ireland[1]

Instrumented Impact Testing for Evaluating End-Use Performance

REFERENCE: Ireland, D. R., "**Instrumented Impact Testing for Evaluating End-Use Performance,**" *Physical Testing of Plastics—Correlation with End-Use Performance, ASTM STP 736,* R. E. Evans, Ed., American Society for Testing and Materials, 1981, pp. 45–58.

ABSTRACT: Several applications of the instrumented impact test for evaluations of polymers and fiber reinforced polymers are discussed. These include simple flex bar, falling dart, and end-use component tests. Instrumented impact testing provides a complete record of the load applied to the specimen during the entire impact event. Distinct characteristics of the load record indicate the deformation and fracture process in the specimen. Subsequent analysis permits detailed study of the various fracture modes. Total absorbed energy is simply segmented into that required for initiating failure and that for propagating the fracture. Results of the laboratory impact test can then be evaluated with respect to the specific properties of interest for the product end-use.

KEY WORDS: impact, force, energy, deflection, polymers, composites, component testing

There are many different physical tests available for evaluating a material structural response. These tests can be separated into two groups; those which subject the material to conditions of environment and deformation which are similar to that expected for the end-use, and those where specific measured material response characteristics have been correlated with end-use performance. In addition to economic considerations, the ideal material selection criteria would be based on information obtained from tests in the first of these two groups. Unfortunately, it is nearly impossible to perform physical tests which are models of the end-use conditions. Therefore, a compromise is usually obtained by testing in a manner such that at least some of the end-use characteristics are included.

Material performance under impact conditions is one of the common end-use considerations. Impact testing has traditionally been a reliable, simple, and inexpensive laboratory tool for determination of the fracture energy absorption characteristics of a variety of materials. Yet, throughout its history, extremely useful (often vital) information has gone undetected. Recent developments in testing, identified as Instrumented Impact Techniques

[1]Manager, DYNATUP Products, Santa Barbara, Calif. 93111.

[*1*],[2] have transformed the conventional impact test into a high strain rate deformation test. This technique provides a complete record of the load applied to the specimen during the entire impact event, in addition to the conventional single value of total absorbed energy. Automated electronic data analysis provides graphical records of applied load and energy absorbed as functions of either time or specimen deflection during impact. Impact performance can be evaluated in terms of all characteristics of the fracture process. For example, total absorbed energy is simply segmented into that required for initiating failure, and that for propagating the fracture. Material end-use does not generally require optimum properties for both resistance to initiation and propagation. Laboratory impact testing can now discriminate the specific characteristics of interest for material end-use.

In the past, impact testing has required use of apparatus which often restricts specimen design to configurations which are far removed from the material end-use. For example, the Izod and Charpy tests are based on use of a free swinging pendulum from which the single value of total absorbed energy is obtained by measuring the energy left in the pendulum after completion of the impact fracture. Not only does this put restrictions on the specimen size, but often the specimens must be specially molded, producing a microstructure not at all like that expected for the material in the end-use condition. Because the instrumented impact test is based on direct measurement of applied load, impact testing is no longer restricted to use of only certain types of machines. Moreover, the specimen geometry is not restricted, and the impact specimen often can be the entire end-use structure or sections from that product.

This paper introduces the instrumented impact test technique through brief discussions of a variety of different applications. Included are examples of the relative simplicity of data analysis. The examples include both bend and falling dart penetration types of testing.

Instrumentation

Instrumented impact testing involves a variety of different impact machines and test specimen designs; however, the basic instrumentation is essentially the same for each type of test [*1*]. That is, each requires an impact machine, a load sensor, and a signal display component. The impact machines include both pendulum and drop tower types. The particular machine employed usually depends on what is more readily available, and is not necessarily the optimum choice for dynamic testing. The most commonly used load sensor is obtained by cementing strain gauges to the striker of the impact machine [*2*].

The signal display component generally utilizes a storage oscilloscope from which is obtained a photograph of the raw data record. However, micro-

[2]The italic numbers in brackets refer to the list of references appended to this paper.

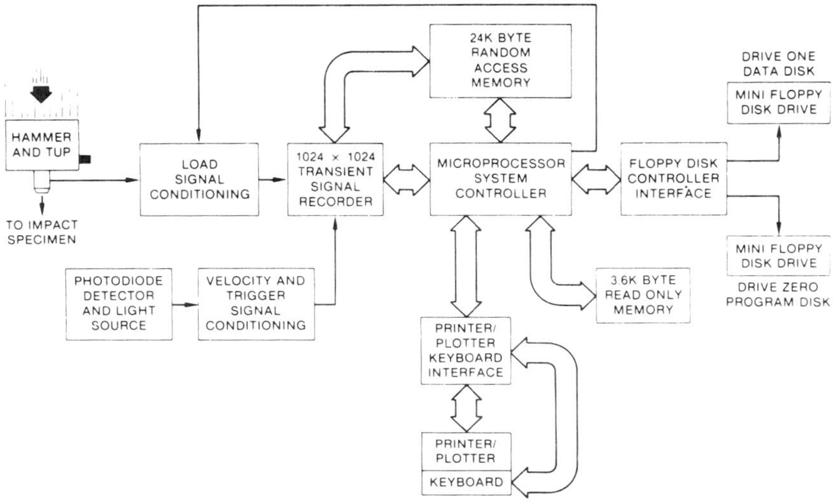

FIG. 1—*Block diagram for automated instrumented impact system.*

processor technology has developed sophisticated automated systems which considerably simplify the instrumented impact test data acquisition and analysis [*3*].

A commercially available automated instrumented impact test includes; direct measurement and computation of pretest parameters, use of transient signal recording methods, microprocessor analysis of raw data, temporary storage of the analyzed results of a series of tests, permanent storage of raw data on a magnetic recording medium (floppy disk recorder), which provides in-depth comprehensive analysis software by floppy disk methods, presentation of analyzed results in tabular and graphical form, and simplified operator interaction requirements through use of a typewriter type of keyboard. The general features of the system are illustrated by the block diagram of Fig. 1, and a more detailed discussion can be found in Ref. *3*.

Data Record

A typical automated instrumented impact test data record is shown in Fig. 2. The graphical presentation of load (P) and energy (E) as a function of deflection (d) is on an approximately 200-mm (8-in.) square plot. The load is measured directly, while energy and deflection are incrementally calculated from the measured velocity (v) at start of impact, known hammer weight, and the load value. The time increments (Δt) for each load value and computation of E and d range from approximately 2 to 8000 μs, depending on the total test time for data collection selected by the operator before the test is run. The relationships employed for these calculations are as follows

$$v_i = v_{i-1} + \left(g - \frac{P_{i-1}}{m}\right) \Delta t$$

$$\bar{v}_i = \tfrac{1}{2}\,(v_i + v_{i-1})$$

$$d_i = d_{i-1} + \bar{v}_i\, \Delta t$$

$$E_i = E_{i-1} + \tfrac{1}{2}m\,(v_i^2 - v_{i-1}^2) + mg\,(\Delta x)$$

where

g = gravitational acceleration constant,
m = impact hammer mass, and
Δx = the distance traveled by the hammer during the time increment Δt.

The term $mg(\Delta x)$ is a correction for gravitational effects during the impact test. That is, for drop weight machines, the energy available for deforming the specimen increases as the hammer continues to drop during the impact. For pendulum machines, this correction is not necessary.

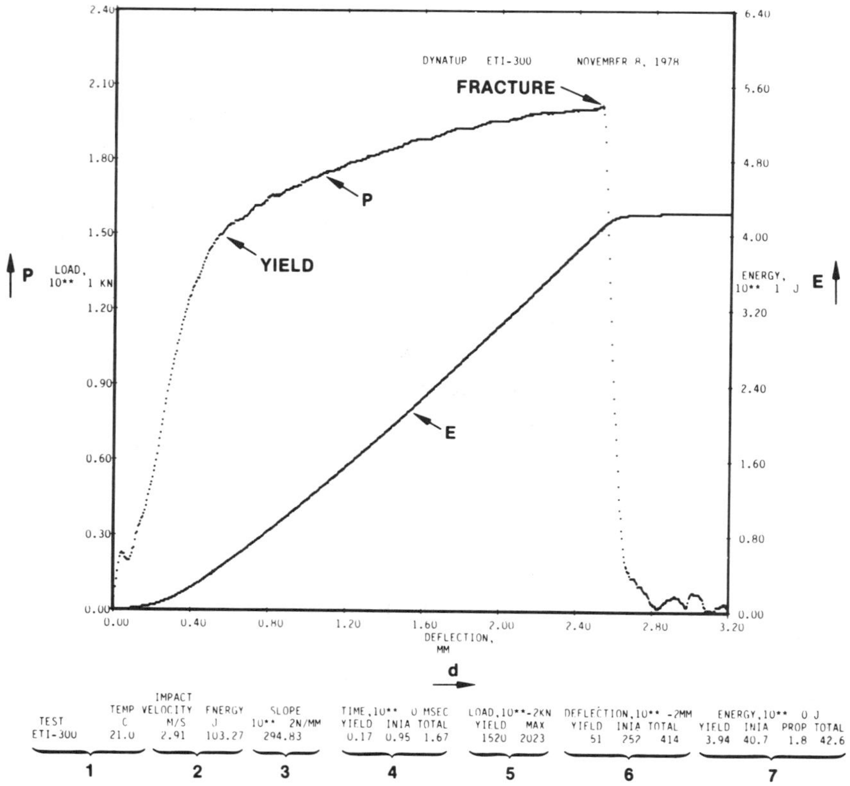

FIG. 2—*Typical data record from automated instrumented impact systems.*

The tabulated data of Fig. 2 include the following:

1. Operator inputed information of specimen identification and test temperature.
2. Measured velocity and the total available energy at impact.
3. Slope of the elastic loading portion of the load-deflection record.
4. Key time parameters at yield and fracture initiation.
5. Key load parameters at yield and maximum load.
6. Key deflection parameters at yield and fracture initiation.
7. Key energy parameters for yield, fracture initiation, propagation, and total.

The data are available for additional analysis (stress, strain, etc.), or for reporting tabulated results of several tests. The graphical presentation of load and time could also be as a function of time during impact. The use of microprocessor technology makes it possible to acquire, analyze, and report complex test results with minimal operator interaction.

Applications and Discussion

The single most important benefit from instrumenting the impact test is the generation of a load record of the entire impact deformation and fracture process. The shape of the load record indicates the stages and type of fracture experienced by the specimen. The record shown in Fig. 2 indicates the specimen exhibited elastic load-deformation behavior up to a point where a yield type phenomenon caused the slope of the load curve to decrease sharply, and then continued to deform until fracture occurred at the point indicated by the sudden decrease in load. The relatively fast decrease in load at fracture indicates a brittle fracture mode.

An illustration of the utility of knowing the shape of the load record is shown in Fig. 3. The specimens were simple flex bars tested in three-point bending. Upon visual examination, both fractures appeared to be brittle with Sample B absorbing approximately four times as much energy as Sample A. The load-time records shown in Fig. 3 relate a clear picture of the actual differences between the two samples. Sample A fractured while the bar was

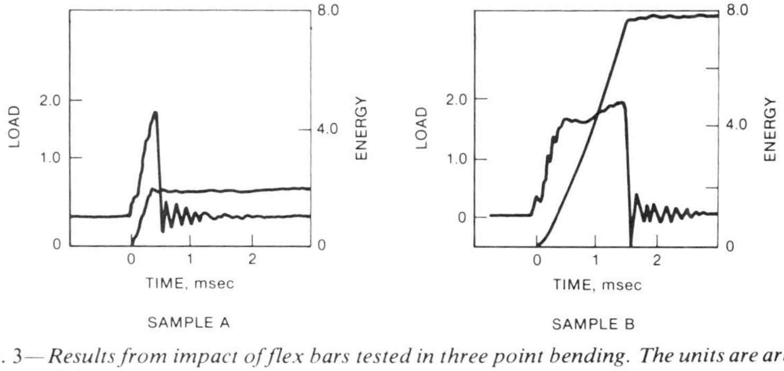

FIG. 3—*Results from impact of flex bars tested in three point bending. The units are arbitrary and intended for comparison only.*

elastically loaded, and Sample B exhibited elastic-plastic deformation before the brittle fracture was initiated. The maximum load or stress characteristics are essentially the same for both samples. However, Sample B exhibited a much larger strain before fracture, as indicated by the longer time to reach fracture initiation.

Any physical property test requires a rational criteria for assessing the relative performance of the material and/or structure. For example, the material end-use does not generally require optimum performance characteristics for both resistance to initiation and propagation of failure. In addition, failure can be either fracture or the elastic limit (yield) point. The illustration shown in Fig. 3 indicates both samples had very little resistance to propagation of fracture. If failure had been defined as the elastic limit, there would be very little difference between these two samples. If failure had been defined as the initiation of fracture, then Sample B would clearly be the better performer.

The best materials evaluation test duplicates as many of the critical product end-use conditions as possible. However, there are exceptions generally occurring with respect to economic considerations. One example is the use of the instrumented impact test to monitor the fatigue life quality of an automotive flex fan. The flex fan is employed for circulating air past the radiator of the engine cooling system.

The flex fan is a clever, materials design utilizing glass fiber reinforced nylon. The centrifugal force imposed by the fan rotation causes the blades of the fan to flex and change contour. At relatively low engine speeds the contour is optimized for circulating air, and at the higher speeds the contour changes so as to have the minimal force required to rotate the fan. That is, normal air convection caused by the vehicle movement at the higher engine speeds is sufficient to cool the radiator. The blade contour change at high speeds assists engine fuel economy.

The material is subjected to fatigue type loading as a result of the flexing to change contour of the blades. Apparently, an excellent test for quality control was to put the fan on an engine in a special test cell, and cycle the engine at high speeds with the surrounding environment controlled at high humidity level. The latter was suspected to be a degrading condition for the nylon matrix. Typical fatigue failures were induced between 80 to 100 h under these extreme conditions. Although this test helped assure a quality, reliable product, the long test times were not compatible with the quality control requirements for making acceptance or rejection decisions on a rapidly produced product.

The instrumented impact test does have the good characteristic of producing results quite rapidly. However, there was no apparent relationship between impact and engine fatigue conditions. Therefore, a research program was initiated to determine if a correlation existed between one or more of the instrumented impact data characteristics and the number of cycles (N) to failure in the engine test.

Small unnotched bend specimens were cut from the leading edge of each of

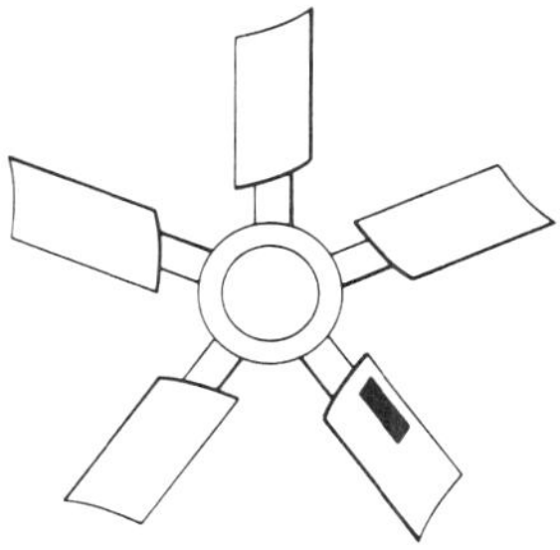

FIG. 4—*Test specimen location in blades of automotive flex fan.*

the five blades of the flex fan, see Fig. 4. Testing was performed with a Charpy pendulum impact machine equipped with an oscilloscope based instrumented impact system.

A typical data record from this type of test is shown in Fig. 5, where

P_{max} = maximum load,

t_f = time to fracture, and

E_t = total absorbed energy.

Each of these parameters was examined to determine the correlation between N and failure in fatigue. A good linear correlation was established between t_f and N. The P_{max} values for various quality fans did not vary significantly. However, the t_f values did exhibit changes which followed the trend of the engine schedule tests. The t_f values are qualitative indicators of the dynamic bulk elastic modulus, which does have a strong influence on the strains induced in the flex fan during operation. Apparently, these strains are the major influence on the fatigue life of the flex fan. It is important to note that the utility of the instrumented impact test for this application was contingent

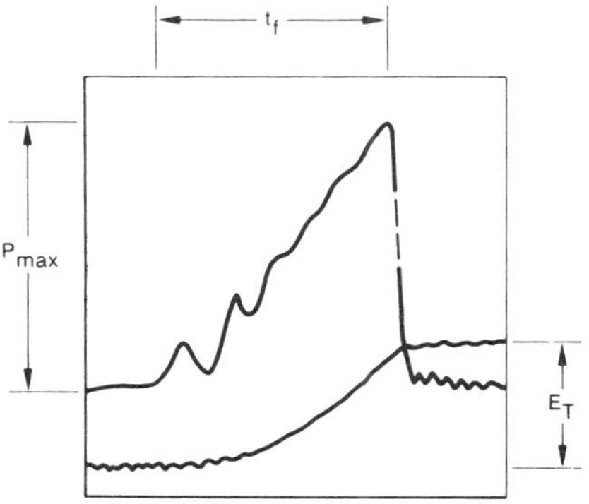

FIG. 5—*Typical data record for flex fan bend specimen.*

upon the empirical correlation, which was established for that specific component. Different materials and/or components would, of course, require a similar correlation to be established before the instrumented impact test could be used as an indicator of fatigue life.

In the resultant quality program, three of the five specimens obtained from each of the flex fan samples are tested, and the average t_f value is used to make the acceptance or rejection decision for that lot of fans. The remaining two —bend specimens are retained for future testing as the need requires. The instrumented impact test became the better test, because of the reduced testing time and the established correlation between t_f and end-use fatigue life for that specific product material.

Fiber reinforced polymers are being effectively used in a number of structural applications. Many times the end-use product will be subjected to impact damage for which a visual examination of the surface impacted does not reveal the extent of damage. One example is the graphite epoxy panels located on the outer surface of advanced aircraft. During servicing or even assembly of the vehicle, a tool may be accidentally dropped or a workman may step on the panel. A decision must then be made as to the extent of damage. A recent test program utilizing the instrumented impact and falling dart impact test methods resulted in the development of a rational decision making procedure for assessing the extent of damage for various shaped tools and heels impacting these composite structures [*4*].

The test program included falling dart tests which varied from those producing no apparent damage, to complete through penetration. The test record shown in Fig. 6 illustrates the type of information obtained from a test for which the penetrator did not completely penetrate the specimen. A panel type specimen was rigidly supported on a rectangular column, and the impact loading essentially imposed a biaxial stress condition in the center of the panel on the face directly opposite the point of impact. The combination of small dart weight and impact velocity was such that the dart did not completely penetrate the sample.

The various stages of impact are indicated in Fig. 6. Incipient damage was

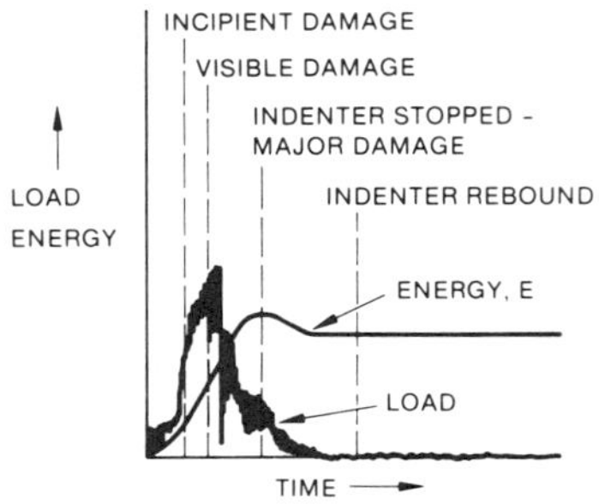

FIG. 6—*Data record for a falling dart impact on a fiber reinforced panel.*

first inferred from the change in slope of the initial elastic loading, and then verified by subsequent special tests. Noted, was the energy absorbed at the point on the load record where incipient damage was suspected. Additional specimens were impacted, with the dart having just the energy suspected to cause incipient damage. Visual and microscopic analysis verified that damage occurred at that energy level. Additional tests at lower dart energy levels revealed no apparent damage.

Tests similar to that described above provided confidence in the interpretation procedure shown in Fig. 6. In most cases the characteristics are clear indicators of the specimen deformation and fracture process. One exception is the point in a low blow impact test when the indenter or dart begins the rebound. For this case, the energy curve has characteristics which clearly indicate the deformation behavior. The energy-time curve is a record of the energy lost by the dart, which is assumed to be that absorbed by the specimen. This absorbed energy is the sum of elastic stored components and permanent deformation components (fracture or plasticity, for example). The peak maximum energy absorbed occurs when all the available energy of the dart has been absorbed through fracture and elastic deformation. This is the point in time when rebound begins. The energy difference between the peak value and the plateau after complete unloading, is the stored elastic energy available for rebounding the dart.

Tests with various shaped indenters, available impact energies, and impact positions relative to panel back-up supports, provided data for developing a rational procedure for estimating the amount of damage to be expected for various end-use conditions. The program also included full–scale fatigue tests to determine the effect of various types of damage on the life expectancy of the panels when in service. The instrumented impact tests provided a method for quantifying the type of damage produced.

Dart impact tests like the ASTM Tests for Impact Resistance of Polyethylene Film by the Free Falling Dart Method (D 1709) and Impact Resistance of Rigid Plastic Sheeting or Parts by Means of a TUP (falling weight) (D 3029) require a trial and error procedure for determining the available dart energy which will initiate failure on the tension surface of the specimen. These methods require substantial numbers of specimens, and the resultant data scatter can be discomforting. It is frequently difficult to separate sample–to–sample material variations from experimental technique variances. In addition, there are some fundamental characteristics of these dart tests which are contrary to the usual considerations given to other tests for these types of materials. For example, the effective strain rate to which the material is subjected in a dart test varies from a relatively high value at the beginning of impact, to a very low value at the initiation of failure. That is, the available energy at impact is carefully adjusted to the level so that just enough is available to initiate cracking or tearing on the tension surface of the sample. This requires the dart velocity to decrease to a zero value near the fracture initiation point. These test characteristics are necessary when instrumentation

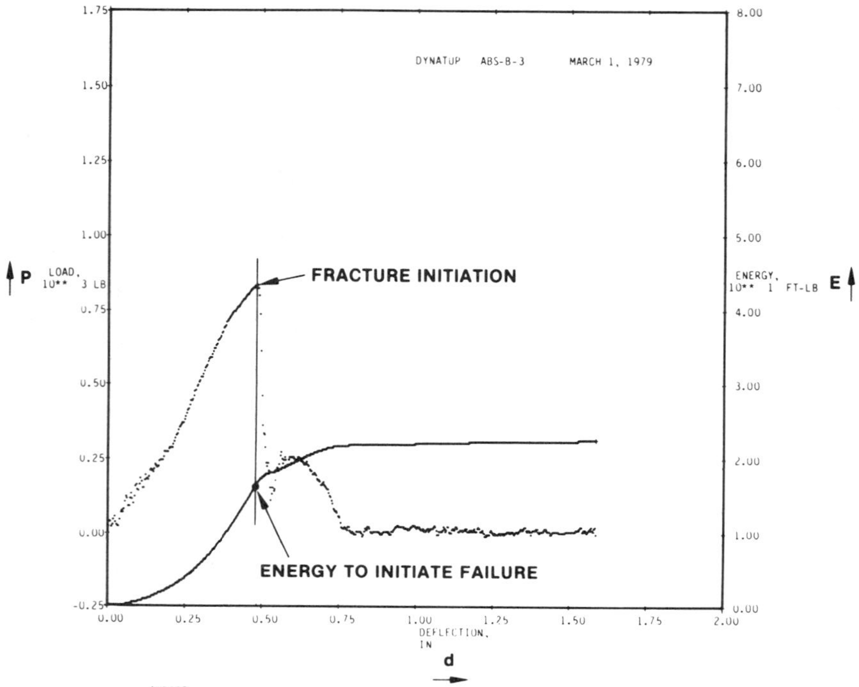

TEST	TEMP F	IMPACT VELOCITY FT/S	ENERGY FT-LB	SLOPE 10** 1LB/IN	TIME,10** 1 MSEC YIELD	INIA	TOTAL	LOAD,10**-1LB YIELD	MAX	DEFLECTION,10** -2IN YIELD	INIA	TOTAL	ENERGY,10** 0 FT-LB YIELD	INIA	PROP	TOTAL
ABS-B-3	71.9	14.39	85.65	179.71	0.28	0.28	1.01	8661	8339	48	48	158	16.29	16.2	6.2	22.5

FIG. 7—*Full penetration data record for ABS sheet, brittle fracture.*

is not available to reveal the total deformation/fracture process during impact.

With an instrumented impact system, an excessive amount of energy can be available to provide complete penetration of the sample. The resultant load record reveals the specimen deformation process for conditions of near constant strain rate. The various characteristics of the load record indicate the initiation of failure and from the attendant energy curve reveals the energy required to initiate the failure.

An example of this full penetration type of data record is shown in Fig. 7. The sample was essentially elastically loaded until failure initiated the sudden decrease in load to a near zero value. The energy to initiate failure is readily determined from the energy curve at the same deflection point for which the load record indicates failure was initiated. The total energy available at impact was 116.2 J (85.7 ft · lb), and the energy required to initiate failure was 22 J (16.2 ft · lb). Additional tests could then determine the relative material property scatter. However, the number of replicates required to determine a representative average failure energy value would, in general, be less than those for the conventional ASTM Method D 1709 or D 3029.

The instrumented impact data also suggest that serious consideration be given to redefining the sample behavior which is to be interpreted as material failure during a dart test. The data record shown in Fig. 8 was obtained from a replicate test of the ABS sheet material employed for that shown in Fig. 7. The data records are quite similar at maximum load. At deflection point 1 in Fig. 8, the total of 24.4 J (18 ft · lb) was absorbed, and compares favorably to the similar point on the record in Fig. 7, where the sample absorbed 22 J (16.2 ft · lb).

The record in Fig. 8 indicates a significant load reduction before the onset of rapid fracture. This relatively slow decrease in load before fast fracture is believed to be indicative of localized thinning of the sample beneath the dart, and not the result of cracking or tearing, which would be interpreted as failure by Methods D 107 or D 3029.

It is suggested that the maximum load bearing characteristics such as load or energy at maximum load, be used to define the impact performance of film or rigid sheet in a dart impact test. If this suggested criteria were acceptable, it would be a simple task to have a through penetration apparatus with digital

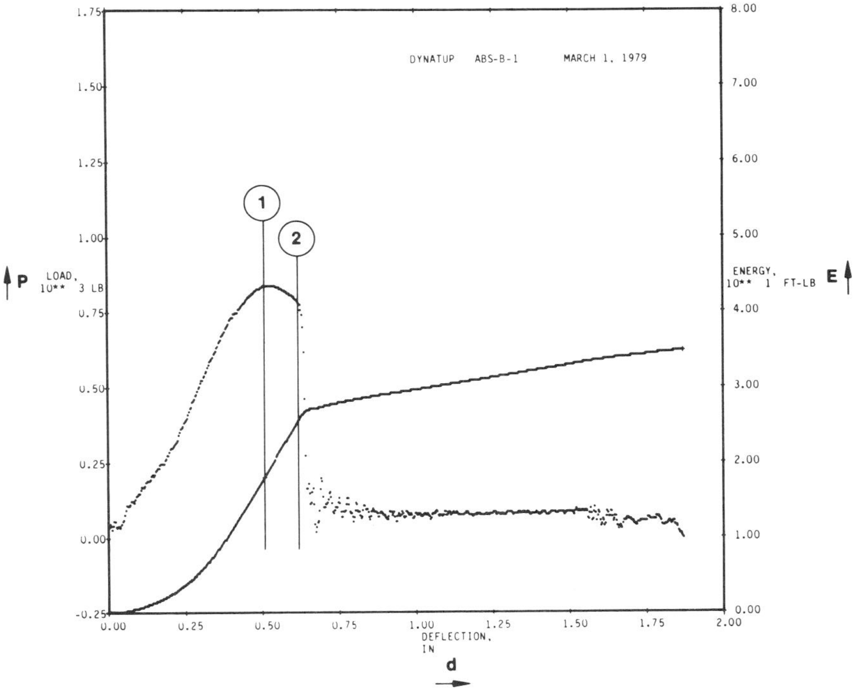

TEST	TEMP F	IMPACT VELOCITY FT/S	ENERGY FT-LB	SLOPE 10** 1LB/IN	TIME,10** 1 MSEC YIELD	INIA	TOTAL	LOAD,10**-1LB YIELD	MAX	DEFLECTION,10** -2IN YIELD	INIA	TOTAL	ENERGY,10** 0 FT-LB YIELD	INIA	PROP	TOTAL
ABS-B-1	71.9	14.70	89.35	175.47	0.29	0.29	1.22	8969	8378	51	51	187	18.05	18.0	16.8	34.8

FIG. 8—*Full penetration data record for ABS sheet, ductile thinning before brittle fracture.*

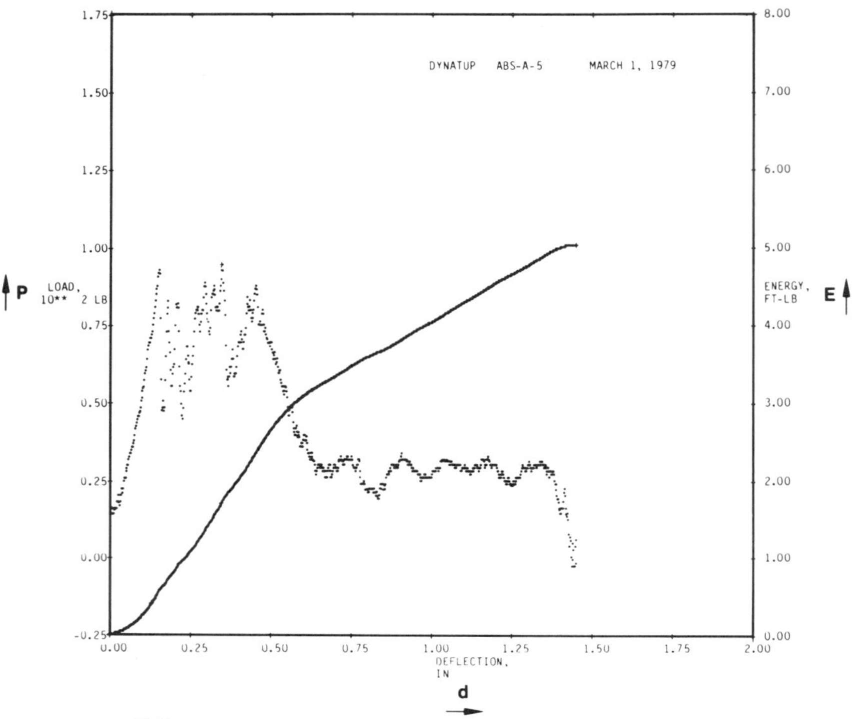

TEST	TEMP F	IMPACT VELOCITY FT/S	ENERGY FT-LB	SLOPE 10** 0LB/IN	TIME,10** 1 MSEC YIELD	INIA	TOTAL	LOAD,10**-2LB YIELD	MAX	DEFLECTION,10** -2IN YIELD	INIA	TOTAL	ENERGY,10** -1 FT-LB YIELD	INIA	PROP	TOTAL
ABS-A-5	71.9	8.54	14.06	558.34	0.12	0.34	1.55	7246	9472	12	34	145	4.00	17.2	33.1	50.4

FIG. 9—*Record indicative of multiple delamination fracture during a full penetration test of ABS sheet.*

readout of the key load and energy parameters. Additional study is required to verify the utility of this approach, but the results to date are encouraging.

The load characteristics are indicative of the type fracture. The record shown in Fig. 9 is for a full penetration test of an ABS sheet specimen which exhibited a delamination type of fracture. The first sudden load drop occurred while the sample was essentially elastically loaded. The subsequent reloading and unloading characteristics are indicative of the major delaminations occurring during the dart penetration.

Polymers have the unfortunate characteristic of being sensitive to the manner in which they are formed in the end-use product. That is, it is difficult, if not impossible to produce molded laboratory test specimens which will have microstructural characteristics representative of the end-use product. In addition, the sectioning of a finished product to make laboratory test specimens, introduces changes in the material, edge affects, for example, which render the specimen questionable as representative of the end-use structure.

All mechanical response tests are models with hopefully a clear correlation or relationship to the final end-use conditions. The best specimen design is the end-use product. Unfortunately, most products do not lend themselves to being used in conventional dart drop or impact bend (Izod, Charpy) tests.

The instrumented impact test reveals the actual load-deformation and absorbed energy-deformation characteristics. This method is well suited to tests of end-use products. Large drop weight machines are commercially available, and when equipped with appropriate instrumentation, provide an excellent apparatus for impact evaluation of the end-use product. Products such as protective head gear, chain saw housings, and drill motor housings are easily tested with instrumented apparatus, see Fig. 10. For these end-use

FIG. 10—*Large weight drop apparatus for instrumented impact tests of football protective head gear.*

structure tests, the load record is used to indicate the deformation and fracture process. The products can be completely destroyed, but the load and energy records during impact clearly indicate the parameters required to initiate the failure.

Conclusions

Instrumentation is available for obtaining complete records of the load-deformation and absorbed energy-deformation behavior of the material during impact testing. Nearly any type of impact test can be instrumented.

The load record indicates the deformation and fracture process, and permits determination of fracture initiation parameters even when the sample is completely fractured by the test.

It has been suggested that the load and energy absorbed at the maximum or limit conditions be used to indicate the impact performance of film and sheet in dart drop types of tests. Additional work is required to verify this suggestion, but the objective would be a simple digital readout device for quality control applications.

Instrumented impact tests have been successfully employed to establish a correlation, and then used as quality control indicators of end-use product fatigue life.

References

[1] Ireland, D. R., *Instrumented Impact Testing, ASTM STP 563,* American Society for Testing and Materials, 1974, pp. 3-29.

[2] Ireland, D. R., "Critical Review of Instrumented Impact Testing," *Dynamic Fracture Toughness,* The Welding Institute, London, July 1976.

[3] Wogulis, S. G. Whitney, B. W., and Ireland, D. R., "Automated Data Acquisition and Analysis for the Instrumented Impact Test," Technical Report No. 78-53, Effects Technology, Inc. 1978.

[4] Aleszka, J. C., *Journal of Testing and Evaluation,* Vol. 6, No. 3, 1978, pp. 202-210.

J. K. Rieke[1]

Instrumented Impact Measurements on Some Polymers

REFERENCE: Rieke, J. K., "**Instrumented Impact Measurements on Some Polymers,**" *Physical Testing of Plastics—Correlation with End-Use Performance, ASTM STP 736,* R. E. Evans, Ed., American Society for Testing and Materials, 1981, pp. 59–76.

ABSTRACT: Instrumented impact of polymeric materials as a function of temperature and in sheet form is a useful tool for characterizing the toughness behavior of the materials in end-use applications.

The methods employed are fast and provide more information than do the classical determinations such as the Izod and Gardner impact tests. Examples of how the load-time and energy-time traces can be used in conjunction with temperature studies to determine ductile-brittle transition behavior will be illustrated.

In addition, the effect of fabrication and flow patterns on impact response and the effect of specimen thickness on impact are discussed.

KEY WORDS: impact properties (instrumented), poly (vinyl chloride), poly (acrylonitrile-butadiene-styrene) (ABS), alloy (polymeric), polymer blends, temperature (effect of on impact), thickness (effect of on impact), ductile-brittle transition temperature, method, toughness

The capability of plastics to withstand an impulsive load without fracture is one of the more important properties of these materials. However, in spite of its practical and technical importance, impact resistance is an ill-defined property generally associated with energy to break in a number of standardized tests normally measured at room temperature. Often the results of these various tests do not correlate well with each other, and materials which perform well in one test do not necessarily perform well in another. This is understandable if one recognizes that the impact resistance of a plastic is not only a function of several material properties but also a function of geometry and fabrication conditions, that is, it is a property of the specimen. To illustrate these points in more detail, this paper is divided into three principal sections: (1) the effect of temperature on impact response of formulated poly(vinyl chloride) sheet, (2) the effect of fabrication on two polymeric alloys, and (3) the effect of specimen thickness on impact response. The bulk

[1]Senior research specialist, Central Research–The Plastics Laboratory, The Dow Chemical Company, Midland, Mich. 48640.

of this paper relates to Section 1 and the location of a ductile-brittle transition temperature (DBTT) for a number of impact modified poly(vinyl chloride) formulations. This temperature, if determined under conditions similar to those under which the material will be used, along with the impact response data, generated from an instrumented impact apparatus, can be readily used to ascertain whether a material is likely to fail in use and whether or not such failure will be catasrophric (brittle) or ductile. However, in making DBTT measurements, the investigator must always keep in mind that a different DBTT may be obtained with a different test geometry or loading rate. While some investigators have recognized the importance of DBTT measurements, they have generally been ignored by many polymer scientists and engineers. To illustrate the importance of the ductile-brittle transition, consider Horsley's statement of "Do not worry about the numerical value of impact strength under standard test conditions . . . be concerned only with keeping on the "good" side of the ductile-brittle transition."[2] Perhaps one of the principal reasons is the large number of specimens required and the time necessary to perform the large number of tests in some of the classical impact experiments. However, in developing a curve for six to seven test temperatures, we have utilized very little additional time and material than for a single temperature determination using the more classical statistical methods. This method will be described later.

General Experimental

Apparatus

The impact tester utilized in all experiments was a Dynatup 8000 drop tower in which a 65.9 kg[3] (145 lb) weight slides uniformly and reproducibly over two ground and polished rods. The instrument was manufactured by Effects Technology, Santa Barbara, Calif. All the work described here was carried out from a drop height of 30.5 cm (12 in.) and the speed of the tup at impact was repeatedly measured to be between 239 and 244 cm/s (94 and 96 in./s). Figure 1 is an illustration of the drop tower.

Figure 2 is an illustration of the tups, with attached strain gauge bridge and the anvils utilized in this work. Two different systems were utilized, one with a nominal full scale deflection of 455 kg (1000 lb) load and the other with a full scale deflection of 1364 kg (3000 lb) load. The thinner plastic sheets and most Charpy specimens are tested using the more sensitive measuring system to provide greater signal to noise ratios.

Calibration of the measuring circuit for load was done by placing the tup in an Instron universal testing machine and loading to several predetermined loads. The Instron was previously calibrated by deadweight loading. The energy scale was calibrated by utilizing a strain-rate insensitive material, as an aluminum killed 0.0457 cm (0.018 in.) steel sheet, and measuring the area

[2]Horsley, R. A., Applied Polymer Symposium, Vol 17, No. 117, 1971.
[3]Note that original measurements are in English units.

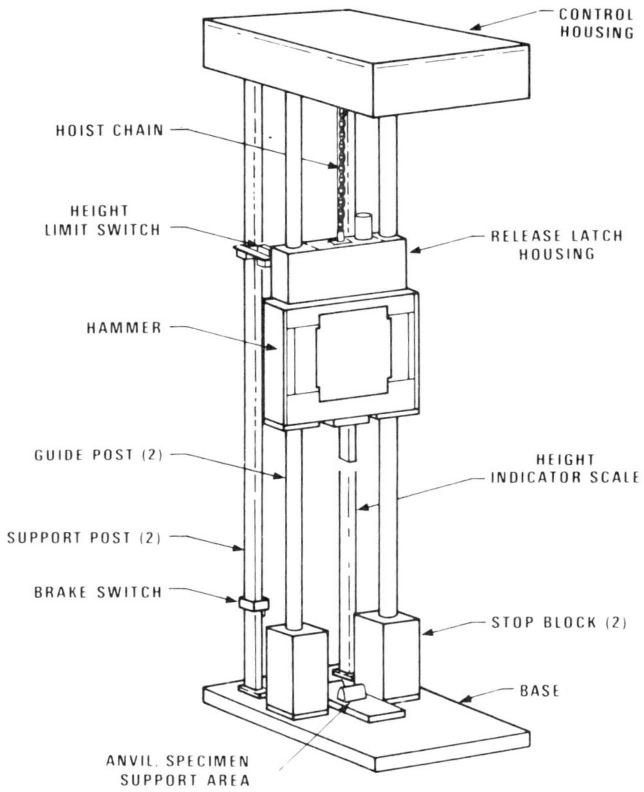

FIG. 1—*Drop tower and components.*

under the stress-strain curve at several different speeds both in the Instron and the Dynatup impact tester using identical geometry in both machines. Both the shape and area under the stress-strain curves were in excellent agreement over the entire strain-rate range. Loading speeds were varied from about 0.0432 cm/s (0.017 in./s) to 355.6 cm/s (140 in./s).

Temperature Measurements

Studies on the effect of temperature on the impact response were all carried out with the cylindrical anvil with a 3.475-cm (1.25-in.) diameter hole and the 0.1588-cm (0.0625-in.) diameter hemispherical tup using 5.08 cm (2 in.) square specimens. The test geometry is similar to that described in ASTM Test for Resistance of Organic Coatings to the Effects of Rapid Deformation (Impact) (D 2794-69) except that the tup is rigidly attached to the drop weight. To determine specimen temperature at impact, a 26 gage thermocouple was embedded in the geometric center of a specimen with the exact shape and thickness of the materials to be tested. Initially a time–temperature calibration curve was obtained by cooling the specimen with the thermo-

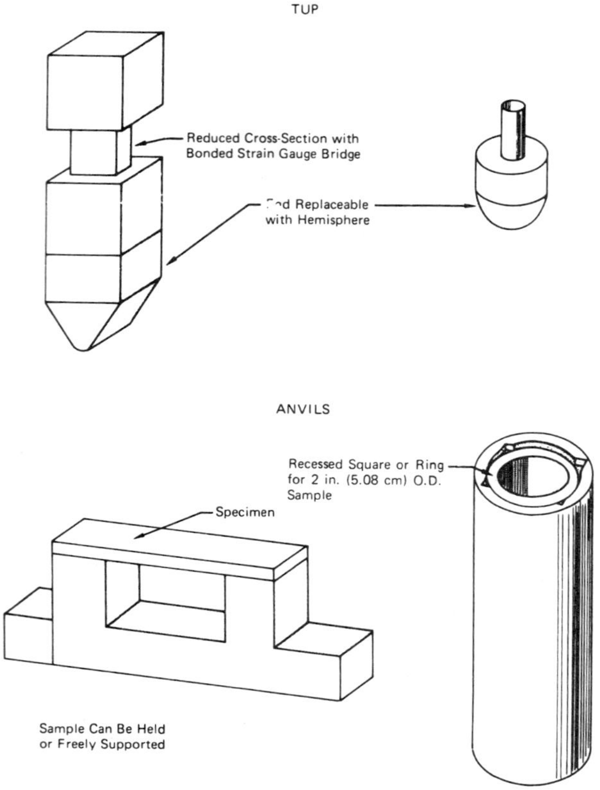

FIG. 2—*Tups and anvils for use with drop tower.*

couple to dry ice temperature and then allowing the specimen to warm up while resting on the anvil. Repeated calibration runs agreed within ±1.7°C (3°F) at any increment of time. After calibration, all specimens to be tested and the specimen containing the thermocouple were placed in a glass reservoir which was immersed in a dry ice bath and cooled until an equilibrium temperature was achieved. Specimens were removed from the reservoir as required, placed on the anvil and impacted after some predetermined time had elapsed. While it is recognized, some temperature gradient exists between the two surfaces of the material, we believe this method is sufficiently accurate for most purposes and it has the great advantage of speed and simplicity.

Impact Data Interpretation

Figure 3 shows representative force–time and energy–time traces (energy and force units are given in the figure captions) for a formulated poly (vinyl chloride) (PVC) extruded sheet of 0.102 cm (0.040 in.) thickness. The specific example used in Fig. 3 consists of 100 parts PVC, 8 parts of a chlorinated polyethylene impact modifier, 10 parts Titanium dioxide (TiO_2) pigment, and about 5 parts of other additives such as heat and light stabilizers, wax, and lubricants.

The trends observed as a function of temperature are generally quite apparent as the impact temperature is lowered from 22.8°C (73°F) to −17.1°C (0°F). In Fig. 3*a*, considerable ductility is observed both before and after the peak force is reached and failure is a ductile tear. Figure 3*b* shows a similar loading curve to Fig. 3*a*, a more rapid decrease in the unloading curve (associated with initial failure which is then arrested) but continues as a ductile tear while the tup continues to penetrate the specimen. In Fig. 3*c* the loading curve is somewhat steeper than in Fig. 3*a* and Fig. 3*b*, reflecting the increase in modulus. Once the failure is initiated, the specimen tends to be quite brittle while is indicated by rapid unloading and examination of the specimens. At −17.1°C (0°F) (Fig. 3*d*) the specimen is almost brittle with only a small level of ductility. Failure is brittle. For most purposes the peak force *(F)*, the energy to peak force ($E_{peak} = E_p$), the total energy to failure ($E_{total} = E_t$), and the difference (ΔE), in E_t and E_p characterize the impact response of the specimen. However, it is sometimes useful to monitor the slope of the loading curve, the time to load and unload, and observe the character of the fracture pattern for unusual features. The methods of data interpretation used here are similar to the techniques described previously.[4,5]

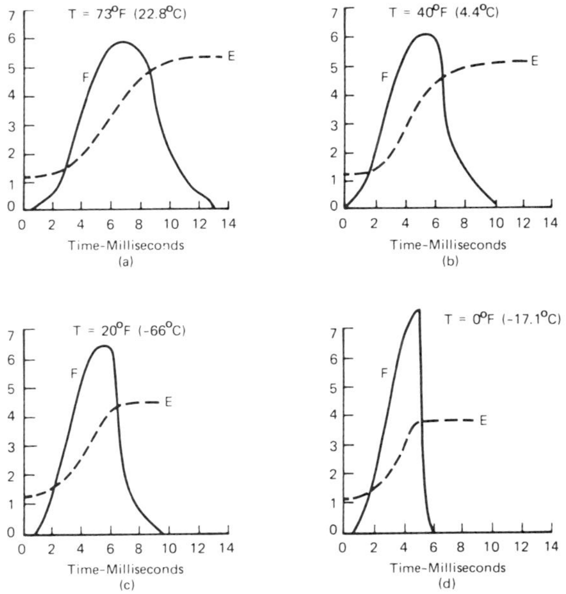

FIG. 3—*Representative force and energy traces for a formulated poly(vinyl chloride) sheet. F = 444.8 N (100 lb)/division; E = 6.78J (5 ft · lb)/division; specimen thickness = 0.102 cm (0.040 in.).*

[4]Lubert, W. Ring, M., and Pavan, A., *Journal of Applied Polymer Science*, Vol 20, 1976, pp. 1107–1116.

[5]Gonzalez, H., Jr., and Stowell, W. J., *Journal of Applied Polymer Science*, 1976, pp. 1389–1404.

Effect of Temperature on Impact Response

In this section, the effect of temperature on the impact response of several different formulated PVC sheet materials is described. All formulations are of similar composition to the one described in the previous section; all were prepared by a similar procedure and differ only in the grade of PVC and the impact modifiers utilized. Formulations were prepared by dry blending and then extruded by a twin screw extruder through a 132.1-cm (54-in.) wide sheet die. Specimens were taken from a portion of the sheet. All sheet samples had the same general appearance and texture when they were received.

Figures 4 to 9 are individual plots (Samples A through F) for the six

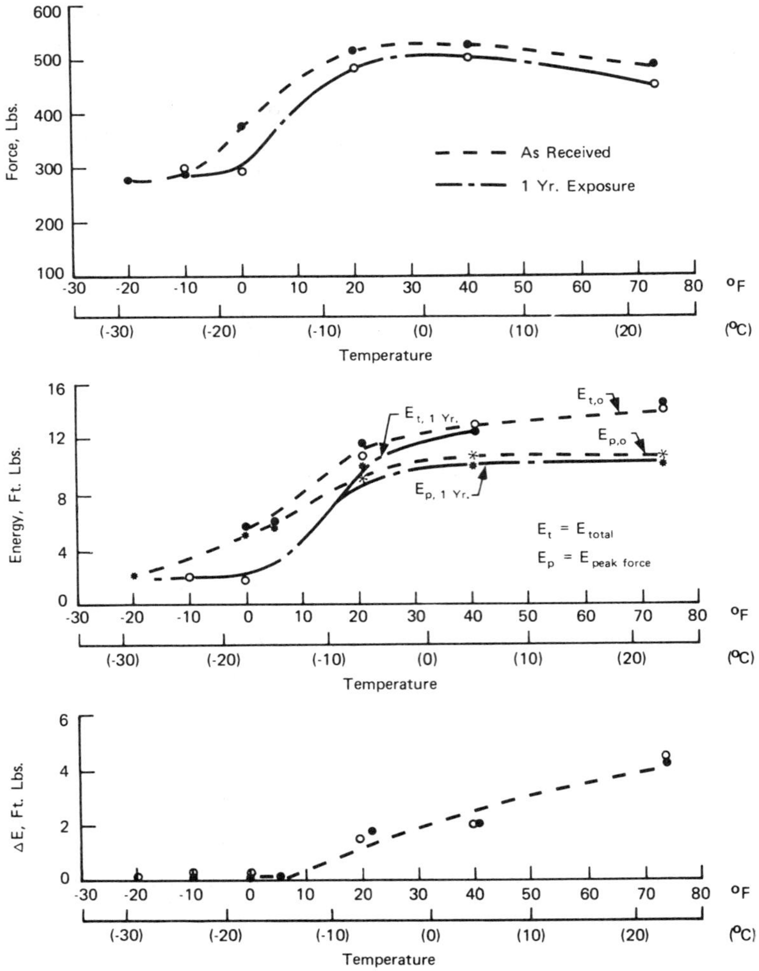

FIG. 4—*Impact response of formulated poly(vinyl chloride) sheet before and after exterior exposure—material A (lb force × 4.448 = N; ft · lb × 1.356 = J).*

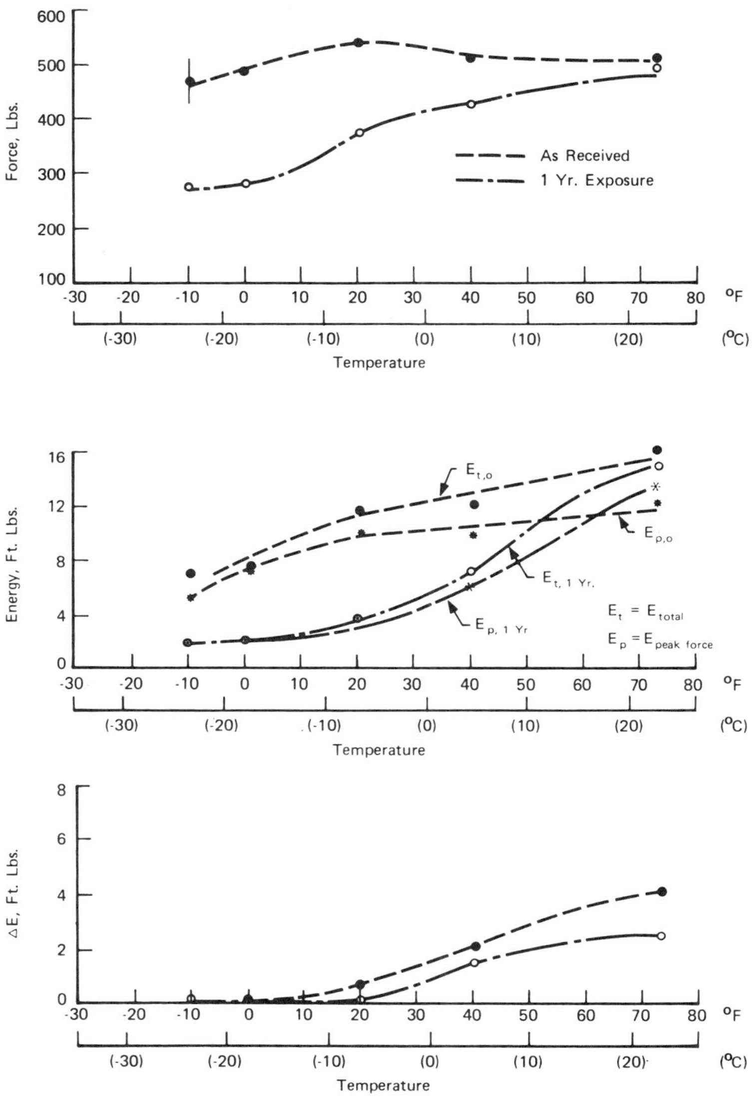

FIG. 5—*Impact response of formulated poly(vinyl chloride) sheet before and after exterior exposure—material B (lb force × 4.448 = N; ft·lb × 1.356 = J).*

different formulations evaluated. The dashed lines represent data generated on unexposed specimens, and the more solid lines are for data generated on specimens exposed for one year to Arizona weather. Each curve is also appropriately labelled. The subscript 0 or 1 indicates whether the samples received exterior exposure. (0 = none, 1 = 1 year). In each case, the peak force (F), energy to peak force (E_p), total energy (E_t), and ΔE (the difference between E_t and E_p) are graphed as a function of temperature. Each data point represents an average of 4 to 5 individual specimens. Careful examination of Figs. 4 to 9 indicates many significant differences between the various formulations (dashed lines), but sample F_0 is clearly superior in low temperature toughness, and sample D_0 is the poorest. The peak in the force–

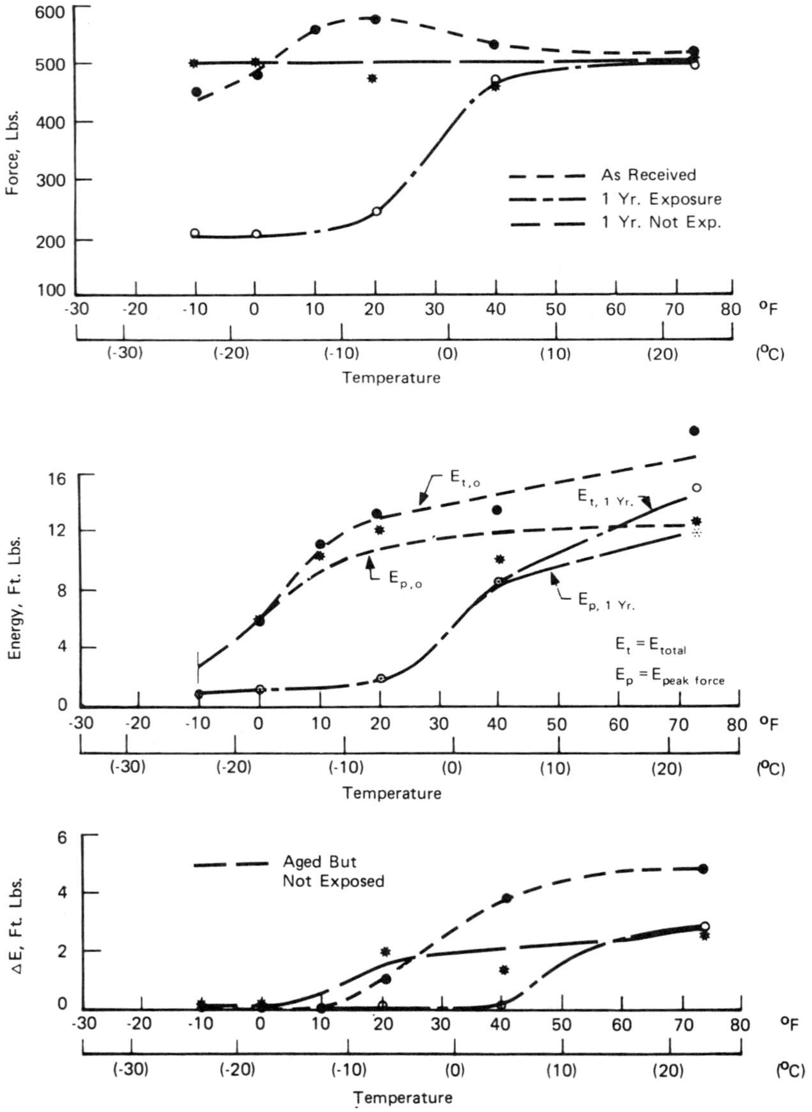

FIG. 6—*Impact response of formulated poly(vinyl chloride) sheet before and after exterior exposure—material C (lb force × 4.448 = N; ft · lb × 1.356 = J).*

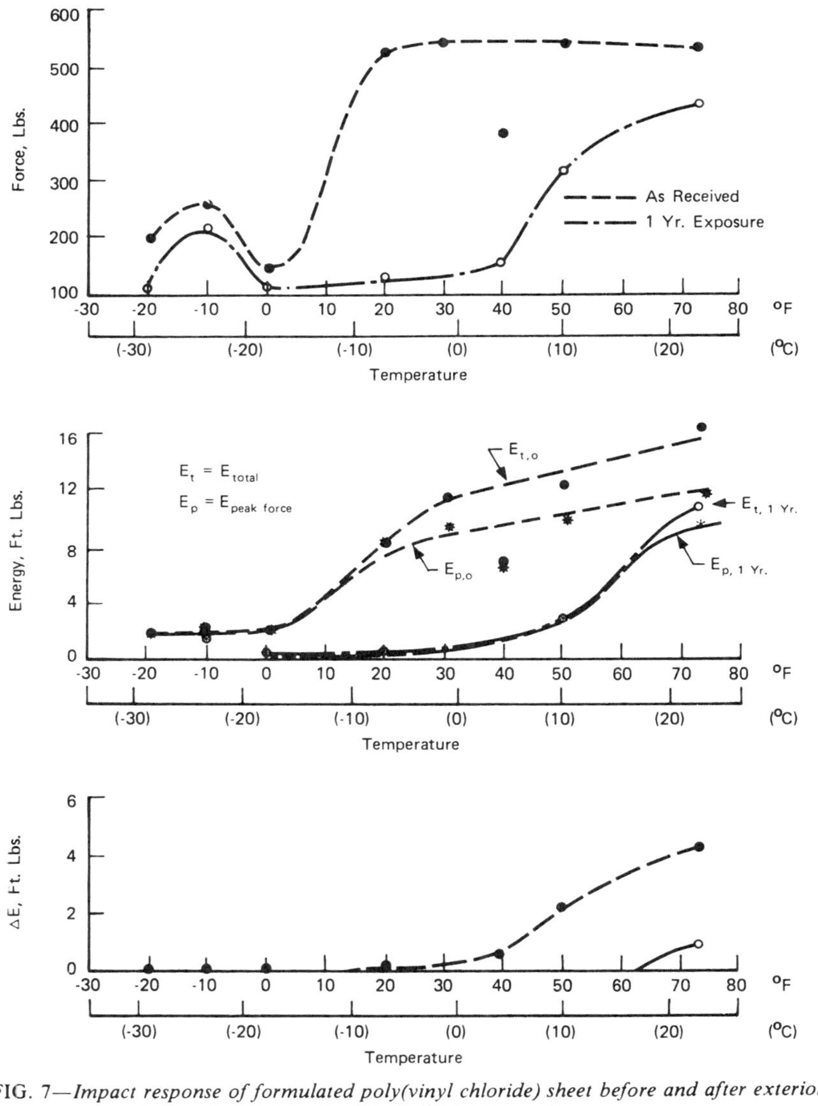

FIG. 7—*Impact response of formulated poly(vinyl chloride) sheet before and after exterior exposure—material D (lb force × 4.448 = N; ft · lb × 1.356 = J).*

temperature relationship shown in Fig. 7 (Material D) is unusual and does not appear in any other system. At present we cannot propose a hypothesis for its existence. DBTT's vary from about −23°C (−10°F) to +9.4°C (15°F). It should be noted that the room temperature impact response would indicate that except for Sample A_0 all materials would perform equally well. (Sample A_0 is about 10 percent lower).

Since these particular PVC formulations were designed for exterior application, samples of sheet were also exposed to sun and weather in Arizona for 1 year. After exposure, the specimens were tested with the exposed side up under the same conditions as used previously. Here again, except for Sample D, very little difference existed between the exposed and unexposed materials when tested at room temperature. Table 1 compares the room temperature impact for all materials.

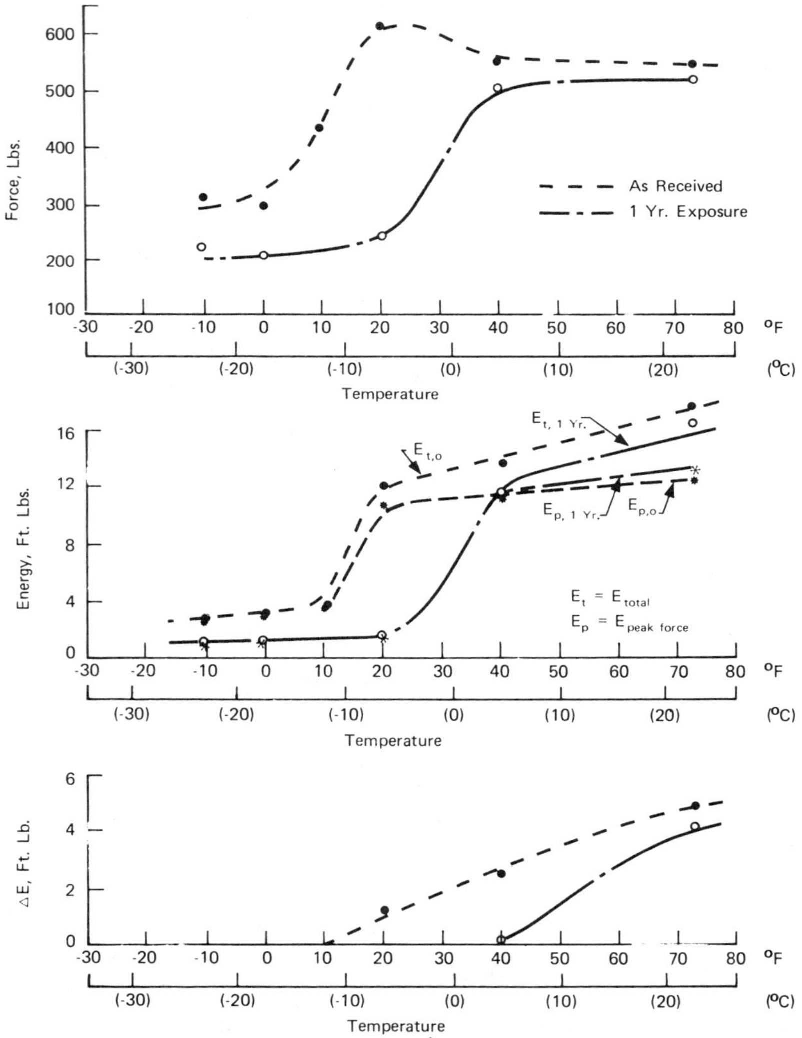

FIG. 8—*Impact response of formulated poly(vinyl chloride) sheet before and after exterior exposure—material E (lb force × 4.448 = N; ft · lb × 1.356 = J).*

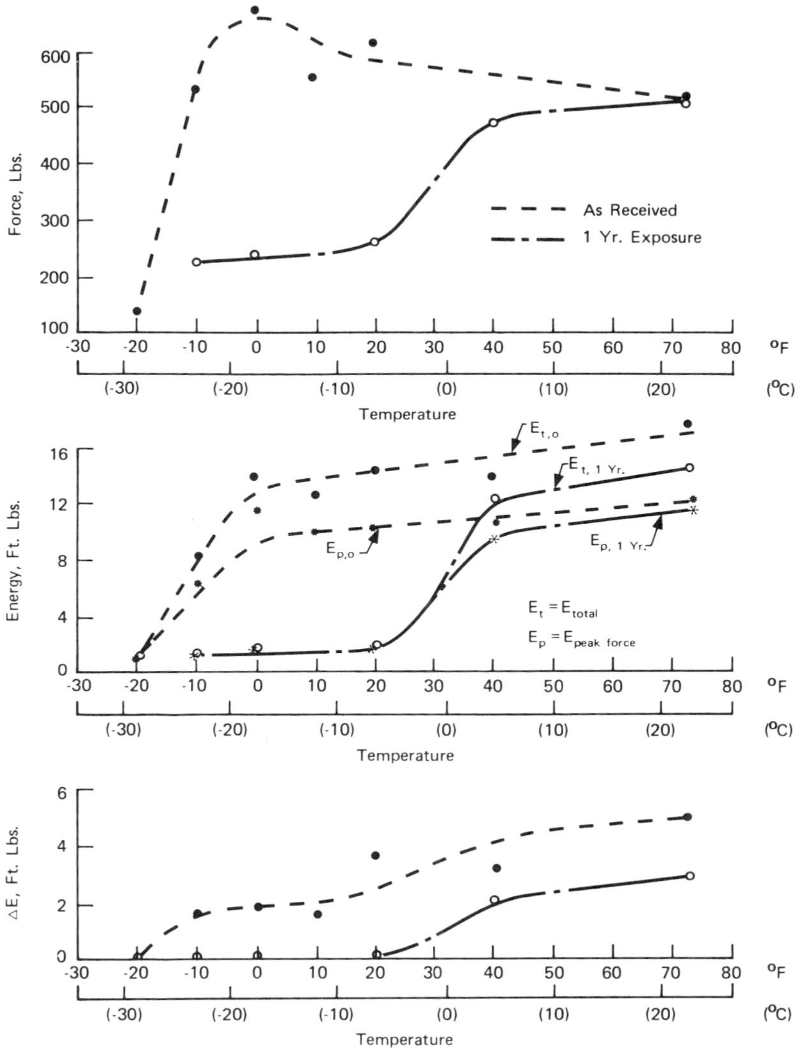

FIG. 9—*Impact response of formulated poly(vinyl chloride) sheet before and after exterior exposure—material F (lb force × 4.448 = N; ft · lb × 1.356 = J).*

The nearly solid lines on Figs. 4 to 9 show the results obtained on the 1 year exposed specimens. The results now are quite different, Sample A has shown almost no change in properties over the entire range of temperatures studied and shows the best impact response at lower temperatures. All other materials have shown substantial shifts in the DBTT. Sample D is still the poorest, followed by Sample B, and then D, E, and F are very similar in performance.

TABLE 1—*Room temperature impact values—initial and 1 year Arizona exposure.*

Sample	Force at Peak, lb	Energy to Peak F, ft·lb	Energy Total, ft·lb	Slope $\Delta f/\Delta t$	$\Delta E = (E_t - E_p)$
A_o	490	10.3	14.5	1.7	4.3
$A_{1\ \text{year}}$	455	10.8	15.3	1.6	4.5
Retain, %	(93)	(104)	(105)		(104)
B_o	520	12.0	16.0	1.7	4.0
$B_{1\ \text{year}}$	500	12.9	15.3	1.7	2.4
Retain, %	(96)	(107)	(96)		(60)
C_o	525	12.4	17.0	1.7	4.6
$C_{1\ \text{year}}$	490	12.3	15.0	1.7	2.8
Retain, %	(93)	(99)	(88)		(60)
C_*	450	11.6	14.5	1.8	2.9
D_o	530	12.3	16.8	1.8	4.5
$D_{1\ \text{year}}$	430	9.6	10.6	1.9	1.0
Retain, %	(81)	(78)	(63)		(22)
E_o	550	12.6	17.4	2.0	4.8
$E_{1\ \text{year}}$	515	12.5	16.5	2.0	4.0
Retain, %	(94)	(99)	(95)		(83)
F_o	520	12.1	17.1	1.6	6.4
$F_{1\ \text{year}}$	500	11.6	14.5	1.7	2.9
Retain, %	(96)	(96)	(85)		(45)

NOTE—retain, % = $\frac{\text{initial value-final value}}{\text{initial value}} \times 100.$

C_* tested about 1.2 year after C_o, not weathered.

lb force × 4.448 = N; ft·lb × 1.356 = J.

In Figs. 10 and 11, we have enlarged the scale and plotted the energy to the peak force (E_p) and total energy (E_t), respectively, against temperature for the exposed materials to better illustrate these points.

At the present time we do not have sufficient supplementary information to indicate or propose a mechanism which satisfactorily explains the shift in DBTT with atmospheric exposure. The exposed sides of the specimen were on the compression side during impact testing and failure is initiated from the tension side of the specimen. The only visible change in the specimens was a very slight amount of yellowing on the exposed side. The observed effects are suggestive of morphological changes occurring; however, we do not have independent confirmation to support this suggestion.

Effect of Fabrication on Impact

To illustrate the effect of flow during fabrication on impact, two different polymer alloys were injection molded into refrigerator crisper trays measuring 27.9 cm (11 in.) by 31.75 cm (12.5 in.) by 8.9 cm (3.5 in.) deep. 5.08 cm (2 in.) square specimens were cut from the trays, identified as to location, and impacted in the Dynatup 8000 using the cylindrical anvil and hemispherical tup. The results are shown in Figs. 12 and 13 as exploded views with each specimen in the same approximate location as it was originally. Each square has enumerated within it the values for peak force (F), E_p, E_t, and the fracture

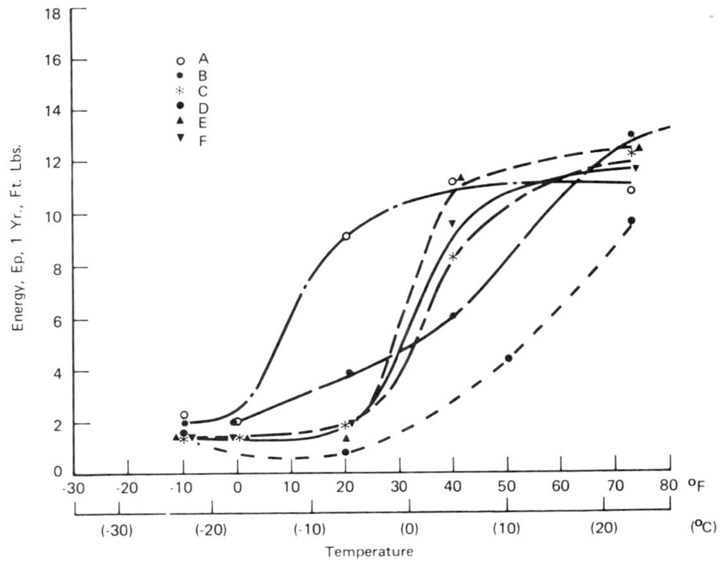

FIG. 10—*Energy absorbed to peak force for one year exterior exposed poly(vinyl chloride) materials (A–F) (ft · lb × 1.356 = J).*

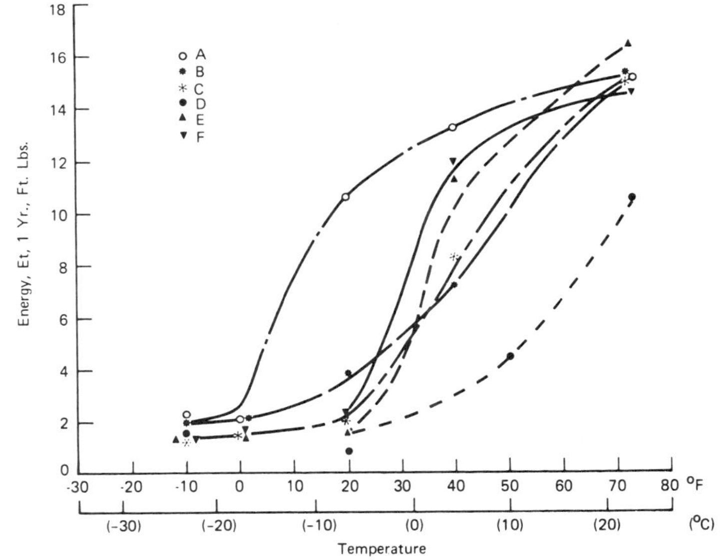

FIG. 11—*Total energy absorbed during impact for one year exterior exposed poly(vinyl chloride) materials (A–F) (ft · lb × 1.356 = J).*

line. In both Figs. 12 and 13, the fracture line is approximately parallel to the flow direction of material. Comparison of results on the bottom of the tray indicate that the two alloys have approximately comparable impact resistance. The average values for Alloy A are F-1544 N (347 lb) and E_t equals 5.56 J (4.1 ft · lb); for Alloy B the equivalent values are 1767 N (397 lb) and 5.61 J (4.14 ft · lb). Tests run on the back and sides of the tray indicate that Alloy B exhibits marked impact improvement (Fig. 13) as it fills the side walls, while the other material is no better and may be slightly worse than results obtained from the tray bottom.

While anisotropy in impact is generally expected with multiphase systems, the use of bi-directional loading, coupled with the force and energy traces, quickly and easily permits an investigator to ascertain where the least tough regions are. Any other test that we have utilized requires many more specimens and time to ascertain the same result.

Effect of Specimen Thickness

In this section of the paper, a high impact acrylonitrile-butadiene styrene (ABS) copolymer was utilized. Samples were prepared by compression

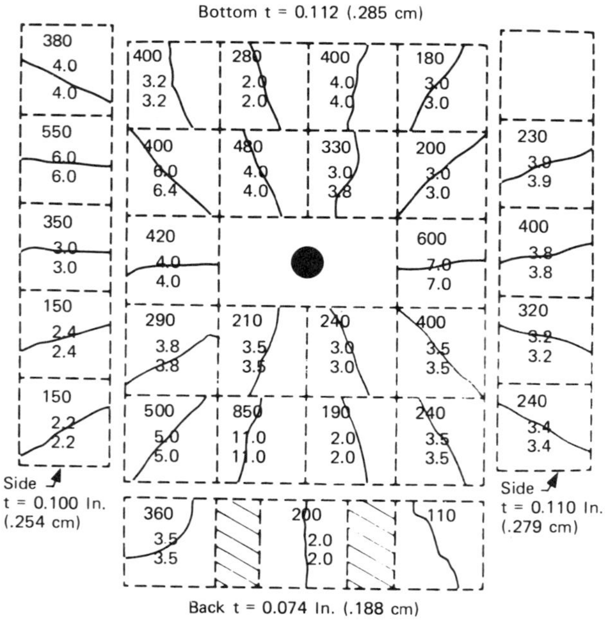

FIG. 12—*Room temperature impact response and observed fracture lines for polymeric alloy A; prepared by melt blending and injection molded into a crisper tray. First no. peak force-lb; second no. energy to peak force ft · lb; third no. total energy ft · lb (lb force × 4.448 = N; ft · lb × 1.356 = J).*

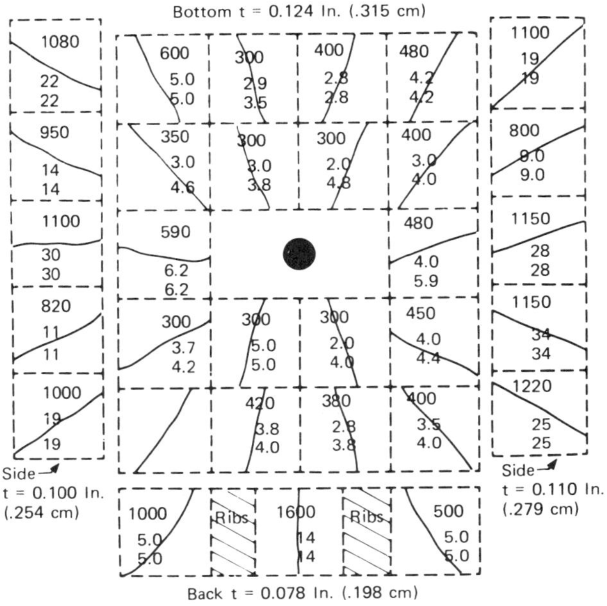

FIG. 13—*Room temperature impact response and observed fracture lines for polymeric alloy B; prepared by melt blending and injection molded into a crisper tray. First no. peak force–lb; second no. energy to peak force–ft·lb; third no. total energy–ft·lb (lb force × 4.448 = N; ft·lb × 1.356 = J).*

molding under time–temperature–pressure conditions such that all materials would exhibit about the same history irrespective of thickness. Five different sample thicknesses were prepared: 0.094 cm (0.037 in.), 0.125 cm (0.049 in.), 0.457 cm (0.180 in.), 0.340 cm (0.134 in.), and 0.490 cm (0.193 in.). After molding, the compression molded sheets were cut into 5.08 cm (2 in.) squares. Specimen to specimen thickness variation was maintained at less than ±0.005 cm (0.002 in.). The 5.08 cm (2 in.) squares were tested using the cylindrical anvil and hemispherical tup as previously described.

Results of these tests are illustrated in Figs. 14 and 15, where F, E_p, E_t, and ΔE are plotted against temperature. Examination of these figures suggests that below −17°C (0°F) all thicknesses have about equal impact response and tend to exhibit brittle failure; between −17°C (0°F) and 22.2°C (72°F) the impact response exhibits a thickness effect. Not only does impact strength increase with thickness, but thicker specimens tend to absorb more energy per unit thickness than thinner materials. For example, at 10°C (50°F) the value is 11.3 J/mm (0.212 ft·lb/mil) for the 0.490-cm (0.193-in.) specimen and 5.55 J/mm (0.104 ft·lb/mil) for the 0.125-cm (0.049-in.) thick specimen. At even higher temperatures these differences become less pronounced; however, more experimentation is needed to firmly establish this later conclusion.

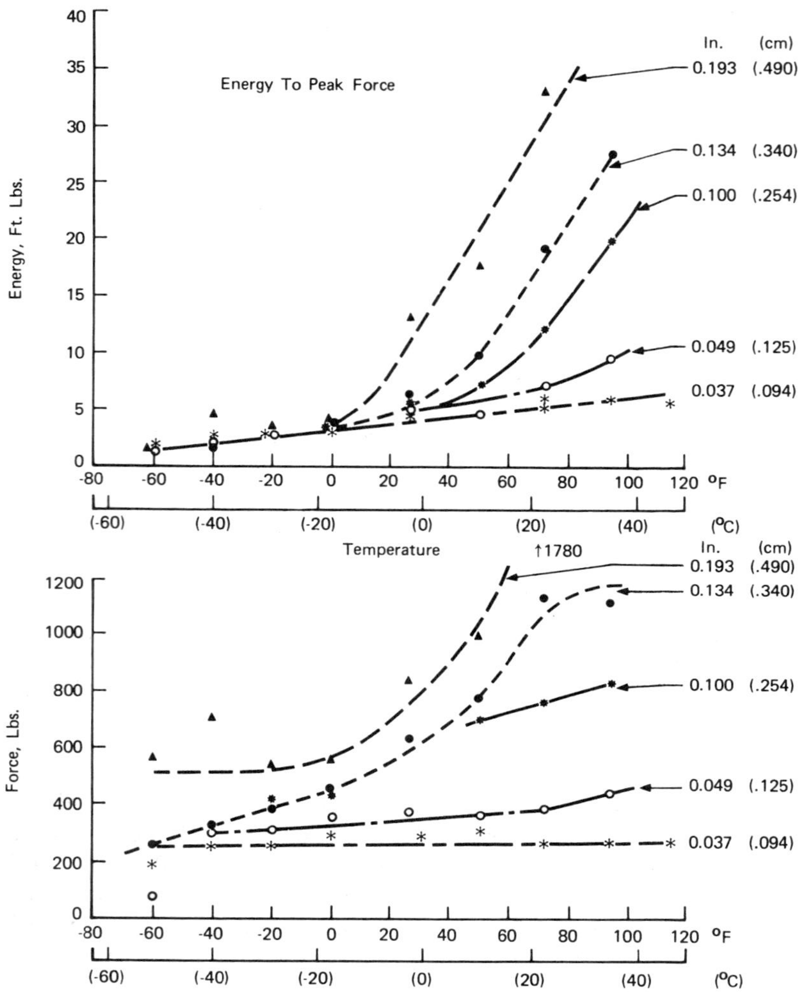

FIG. 14—*Energy to peak force and peak forces reached during impact for compression molded sheets of acrylonitrile-butadiene-styrene copolymer (ABS) for various thicknesses and temperatures (lb force × 4.448 = N; ft·lb × 1.356 = J).*

Conclusions

The impact response of three different material systems has been studied using an instrumented impact tester in a particular geometry. From the data generated we conclude that:

1. Based on our experience, we believe instrumented impact measurements can usually provide more information for the same amount of effort and material than can be obtained by the more classical methods being utilized.

2. The importance of locating a ductile-brittle transition temperature has

been demonstrated for a series of materials. The effect of aging, compositional variables and other external factors can affect the location of this transition temperature. Hence, efforts to relate impact strength to fundamental molecular motions is not always easily accomplished. In many cases such a relationship may be of secondary importance in practical applications

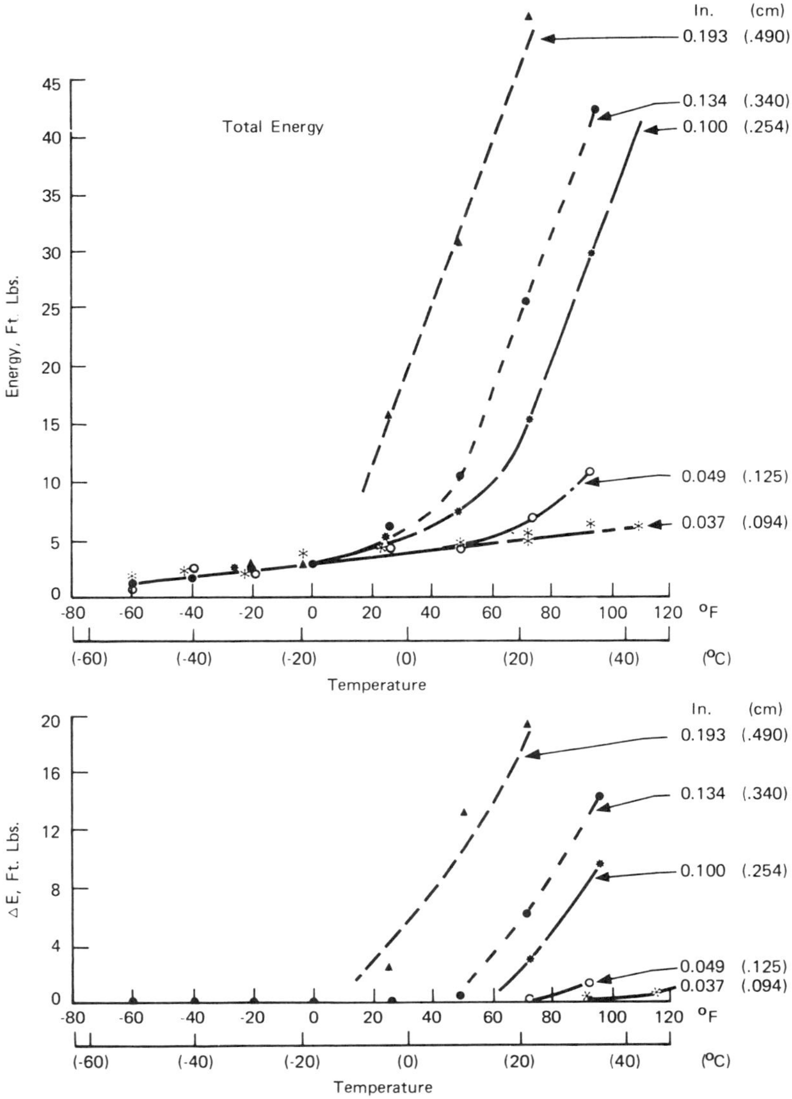

FIG. 15—*Total energy absorbed and* ΔE *calculated during impact for compression molded sheets of acrylonitrile–butadiene–styrene copolymer (ABS) for various thickness and temperatures (lb force* $\times$ *4.448 = N; ft·lb* $\times$ *1.356 = J).*

because of the great number of external variables which influence the response of polymers under impulse loading.

3. The hemispherical tup and the cylindrical geometry of the anvil may be useful in locating weak impact areas in fabricated articles without resorting to cutting and testing specimens at various angles relative to the polymer flow.

4. The strong influence of both thickness and temperature on the impact response of compression molded ABS materials has been demonstrated. Data generated from instrumented impact measurements have facilitated this study. Observations of the force–time and energy–time traces coupled with observations made on failed specimens help in determining whether failure is catasphoric (brittle) or ductile.

Acknowledgments

The author would like to acknowledge the help of V. Coomer and P. Herlihy in generating much of the data. W. Worsley and W. Rupprecht supplied some of the fabricated materials used in this work.

J. F. Carroll[1]

On Abrasion, Scratch, and Friction Testing—Relationship to End-Use Application

REFERENCE: Carroll, J. F., "**On Abrasion, Scratch, and Friction Testing—Relationship to End-Use Application,**" *Physical Testing of Plastics—Correlation with End-Use Performance, ASTM STP 736,* R. E. Evans, Ed., American Society for Testing and Materials, 1981, pp. 77–90.

ABSTRACT: In the field of plastic testing, reproducibility is usually emphasized in standard test methods. However, relevancy is at least equally important to end-use applications. In the field of surface testing, relevancy can sometimes be more difficult to achieve.

Both ASTM and nonstandard abrasion tests for plastics are compared with ANSI abrasion tests for photographic film. Mechanisms include falling carborundum, abrasive loaded wheels, soft-hair wheelbrush, and spherical stylus. All give similar rankings when testing simple plastics surfaces. Only the spherical stylus produces results reliably related to trade experience with the thin coatings on photographic films.

Lubricity is introduced relative to its effect on abrasion resistance and ease of product handling by mechanical equipment. The classical ASTM friction test for plastics is a sled test. A sled test's lack of relevancy to scratch testing exemplifies a breakdown of the classical laws of friction. In the photographic film field, a point-contact friction test is more correlative. The ANSI version of this, the Paper Clip Friction Test, is described, and correlations with scratch as well as with lubricant levels in molded plastics are presented.

KEY WORDS: abrasion, scratch, friction, wax, lubrication, photographic film

In standard test methods, reproducibility is usually emphasized. However, relevancy is at least as important to end-use applications. The fact that a test method is a standard does not assure its relevance in any one specific application. In the field of surface testing, relevancy can be more difficult to achieve. This has proven to be the case in the photographic film field.

Abrasion and Scratch

Travelling films, such as motion picture films and microfilms, are repeatedly subjected to abrasive influences during their use. Most of these films consist of a gelatin layer on one side and a plastic support on the back. In

[1]Technical associate, Eastman Kodak Co., Rochester, N.Y. 14650.

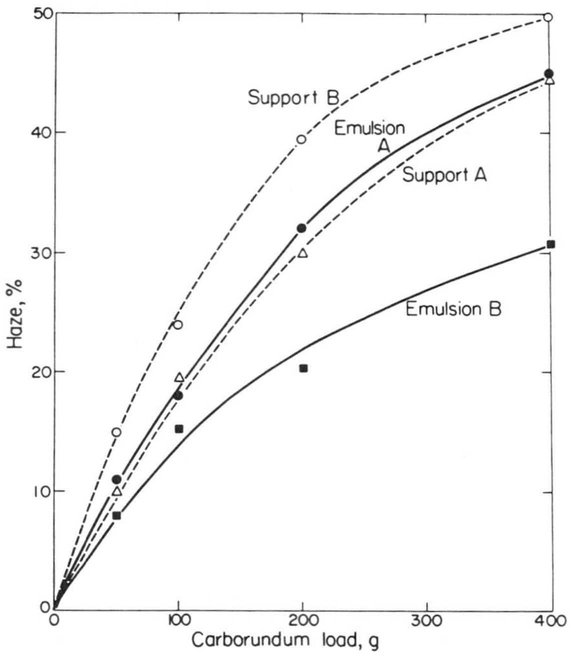

FIG. 1—*Mar resistance of plastics (ASTM D 673-70).*

searching for the most appropriate abrasion test, it was logical to evaluate at first the available American Society for Testing and Materials (ASTM) tests for plastics.

One of the earliest formal abrasion tests applied to photographic film was the ASTM Test for Mar Resistance of Plastics (D 673). Abrasion is produced by a measured quantity of falling carborundum particles and is evaluated by haze measurements. In this method and others in which the area abraded covers the entire evaluation area, the final haze has been corrected for the original haze in the sample by the following formula [*1*].[2] This has been established both empirically and by derivation

$$\% \text{ corrected haze} = \left(\frac{\% \text{ final haze} - \% \text{ initial haze}}{100 - \% \text{ initial haze}}\right)(100)$$

Some examples are shown in Fig. 1 of Method D 673 results for both gelatin emulsions and cellulose acetate supports. Emulsion A and Support A are shown to be quite similar, whereas trade experience has shown that Support A is quite superior to Emulsion A. Therefore, this test method is not applicable to the end-use of these products. This is an example of a standardized test that is not relevant in a given application.

[2]The italic numbers in brackets refer to the list of references appended to this paper.

Another well-established ASTM test in the plastics field is "Resistance of Transparent Plastics to Surface Abrasion" ASTM Test for Resistance of Transparent Plastic Materials to Abrasion (D 1044-78). This test method employs abrasive wheels to produce tracks of measurable haze on a specimen mounted on a turntable. The results are often more relevant to photographic films, but the 100 mm (4-in.) square specimen requirement restricts its use for 8, 16, and 35-mm films.

An example of a nonstandard abrasion test that showed some promise at first was the Soft-Hair Wheelbrush Test. This consisted of a 230 mm (9-in.) diameter horsehair wheelbrush rotating against a specimen mounted on a stationary metal backing. In just a few rotations, a measurable haze was produced that correlated very well with trade experience for the support sides of many films (Fig. 2). However, the emulsion side of certain films (Fig. 3) were severely misranked in comparison with trade behavior. For example, Curve C should be even lower than Curve A. Therefore, this test method also is not applicable to the end-use of these products.

A test method which does correlate with the many facets of trade experience with photographic films is a spherical stylus scratch test. It consists of a counter-balanced arm containing a stylus mount at one end with a weight pan directly over it. Weights are added to load the stylus. A moving specimen holder draws the film along beneath the stationary stylus. The scratches

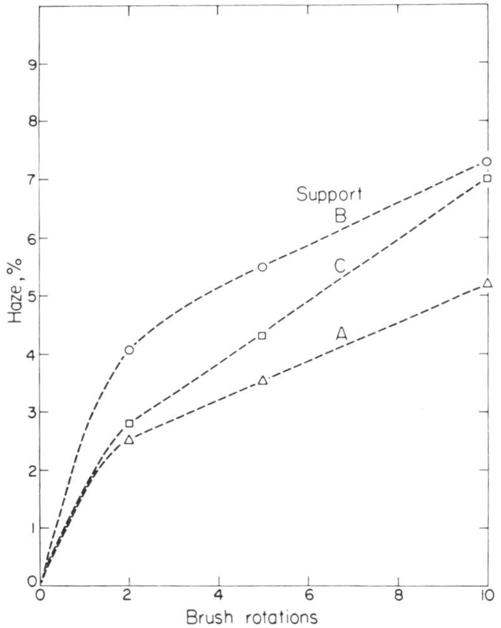

FIG. 2—*Soft hair wheel brush test (processed photographic film supports).*

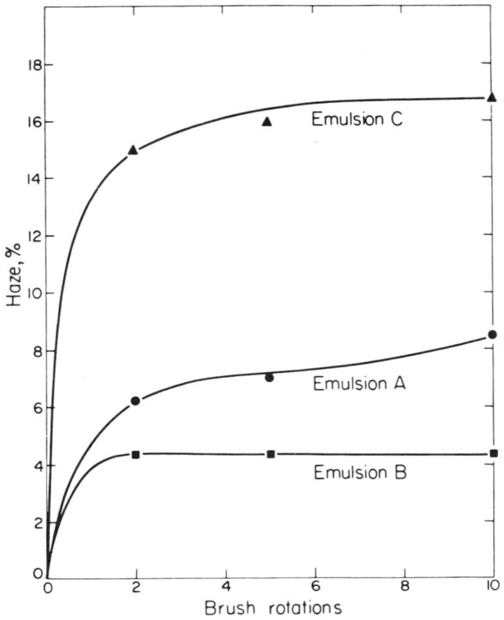

FIG. 3—*Soft hair wheel brush test (processed photographic emulsions).*

produced by the 3-mil radius sapphire stylus are coarser than typical trade scratches, but the test correlates with trade experience nevertheless. This is now the American National Standards Institute (ANSI) test method, "American National Standard Methods for Determining the Scratch Resistance of Processed Photographic Film" (PH1.37), in the photographic standards.

One procedure of this method uses the increase in measured haze as the readout. In this procedure the original haze correction consists of direct subtraction since very little of the measured area is actually abraded. It is interesting to compare the results of this test method (Fig. 4) on Emulsion A and Support A with the rating shown for the same surfaces in Fig. 1. The Fig. 4 rankings are in agreement with trade observation.

Friction and Scratch

If emulsion A (Fig. 4) is waxed in a manner that is typical in the industry, the resistance to scratch is markedly improved (Fig. 5). This is very logical, and it has been well recognized for many years in the motion picture field that low friction is beneficial to scratch resistance [*2–5*]. However, in a typical sled friction test with metal against the emulsion, the difference between a waxed film and an unwaxed film is not very impressive (for example, coefficients of 0.25 versus 0.42). E. Rabinowicz many years ago, however, called attention to

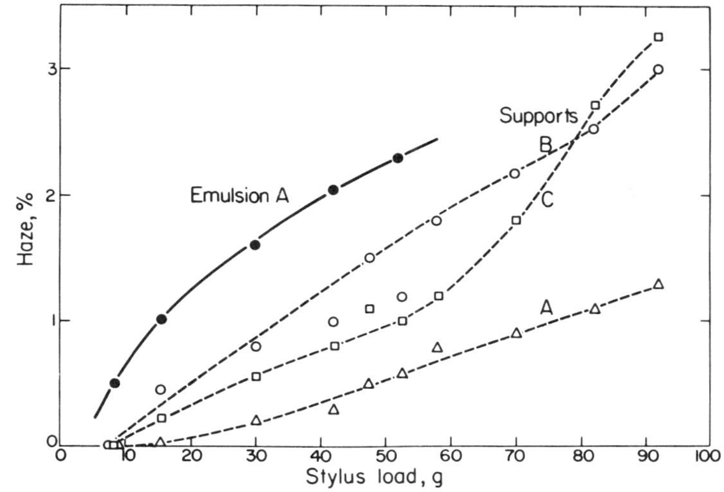

FIG. 4—*Scratch resistance of processed photographic film (ANSI PH 1.37-1977).*

the fact that the coefficient of friction is not independent of unit pressure and real contact area [*6*]. If it is desired to have a friction test to correlate with scratch resistance, then why not one with similar geometry and stress? With this purpose in mind, the Point Contact Friction Test was designed (Fig. 6). Using a series of low wax coverages on the same emulsion surface (Fig. 7), it is evident that the Point Contact Test is more sensitive to lubrication than a metal sled test. Moreover, the shape of the curve for point contact friction

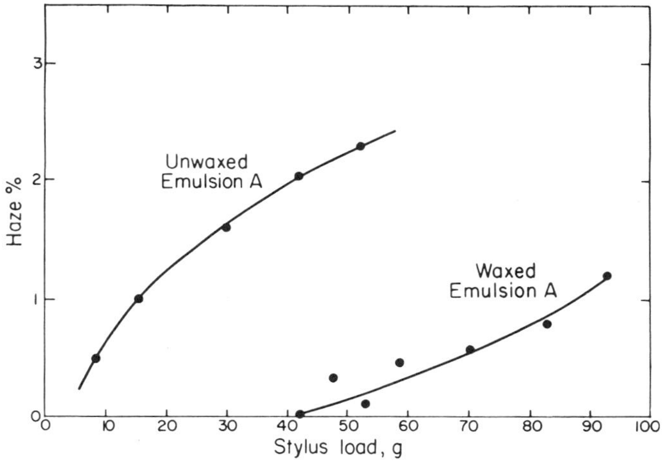

FIG. 5—*Scratch resistance of processed photographic film (ANSI PH 1.37-1977).*

FIG. 6—*Point contact friction tester.*

corresponds more to the shape of the scratch test curve run on the same series of films (Fig. 8). (These scratch ratings are in terms of the stylus load needed to produce a visible scratch, an alternate procedure in ANSI Method PH1.37.) If scratch ratings and friction are compared directly (Fig. 9), a linear relationship is obtained when only the lubricity is changed; that is, the indentability and shear strength are held constant. This relationship permits analysis of the dominant property contributions to scratch ratings. For example, the differences in scratch among the surfaces shown in Fig. 10 would be entirely due to lubricity differences, not inherent differences in polymer integrity, if the points had all fallen along a straight line between the circular points. The fact

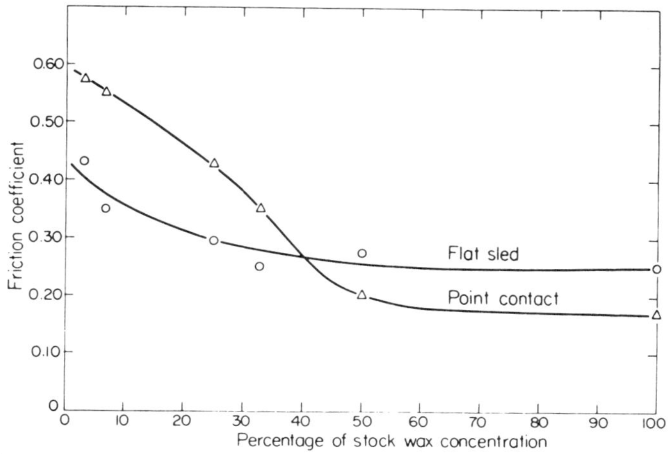

FIG. 7—*Comparison of friction test responses.*

that Points B and C fall to the right of such an imaginary line indicates somewhat greater scratch resistance than can be explained by lubricity alone.

Friction and Projection Performance

An important requirement for satisfactory projection of super–8 films is waxing after processing. Inadequate wax leads to short lifetimes before the onset of unsteadiness or jumpiness of the projected image (Fig. 11). Ultimately permanent film damage occurs with further projection (Fig. 12).

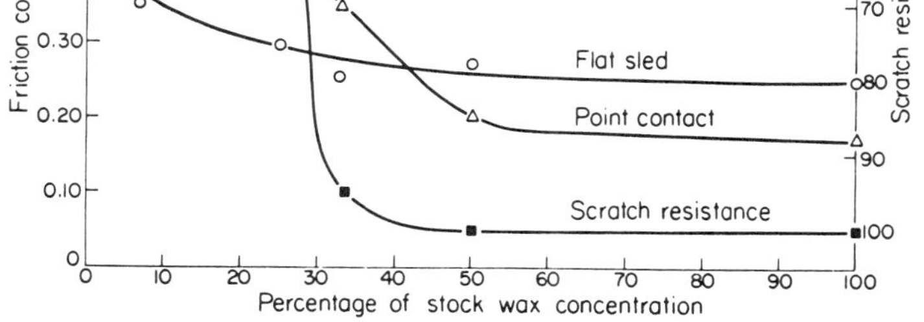

FIG. 8—*Friction responses relative to scratch resistance.*

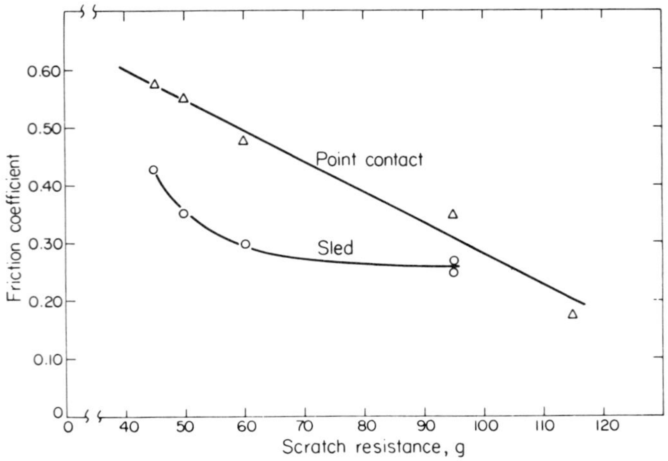

FIG. 9—*Correlation of scratch resistance with two friction tests (film emulsion features lubrication variations).*

A common flat sled friction test is not a sensitive indicator of the adequacy of film lubrication, and until recently a formal chemical analysis was required. Based on the experience of the Point Contact Friction Test, a simple device called the Paper Clip Friction Test was designed to detect film lubricants (Fig. 13) [5]. It is basically an inclined plane friction test with a winged slider whose only means of support is the distal end of a paper clip secured well above the center of gravity. It is very sensitive to lubricants but not to other common

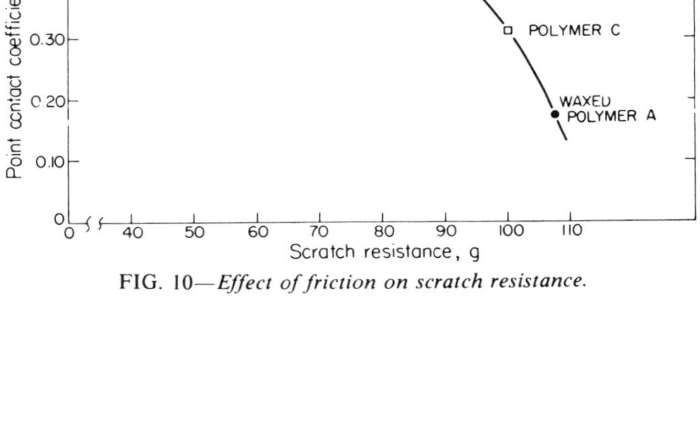

FIG. 10—*Effect of friction on scratch resistance.*

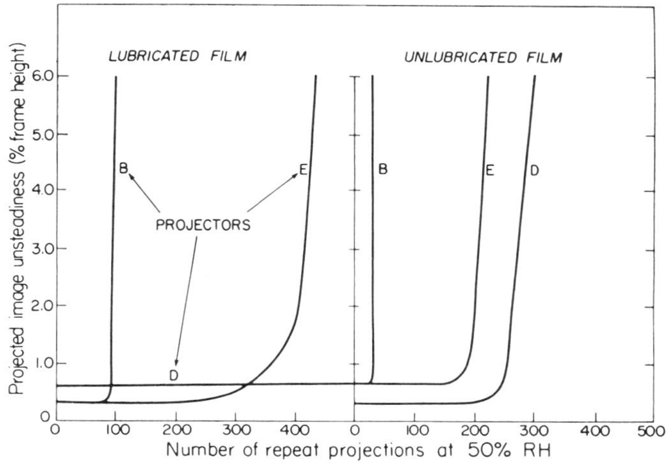

FIG. 11—*Effect of lubrication on unsteadiness in various projectors.*

friction parameters, as shown in Fig. 14 by the contrast in response of this test and a sled test to four different surfaces. This test has the advantages of high sensitivity to lubricants and superior reproducibility; it is self cleaning, and able to operate on minute areas, as, for example, in checking for edge wax on a narrow strip of film. Figure 15 shows the close relationship between the projection life of photographic films and their level of lubrication as quantified by the Paper Clip Friction Test.

This test method has become the "American National Standard Methods for Detecting Lubrication on Processed Photographic Film by the Paper-Clip Friction Test" (PH1.47), and is presently under consideration as an International Standards Organization (ISO) Method in Technical Committee TC-42. In the latter regard, paper clips from five major countries have been tested at three different friction levels (Fig. 16) with astounding agreement despite the different shapes, sizes, radii, platings and surface textures. Apart from photographic film purposes, the Paper Clip Friction Test has been also used successfully to confirm the level of incorporated lubricant in plastics molding experiments (Fig. 17).

Discussion

The important feature in these studies is the requirement that the test method must be pertinent to the end use of the material. While test reproducibility is essential in any procedure, this is not an adequate criterion by itself of test acceptability. A reproducibly wrong result is of limited use.

In the examples shown, the pertinency of the stylus scratch test was

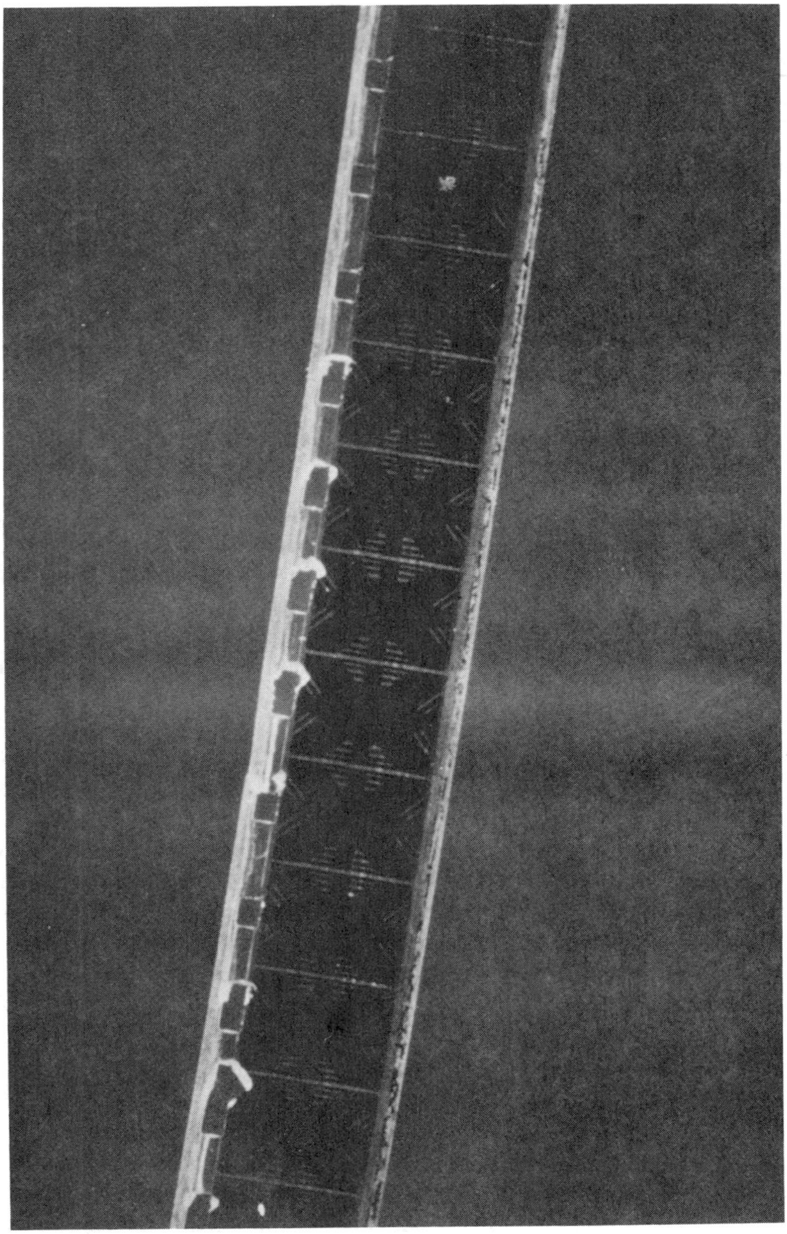

FIG. 12—*Perforation damage.*

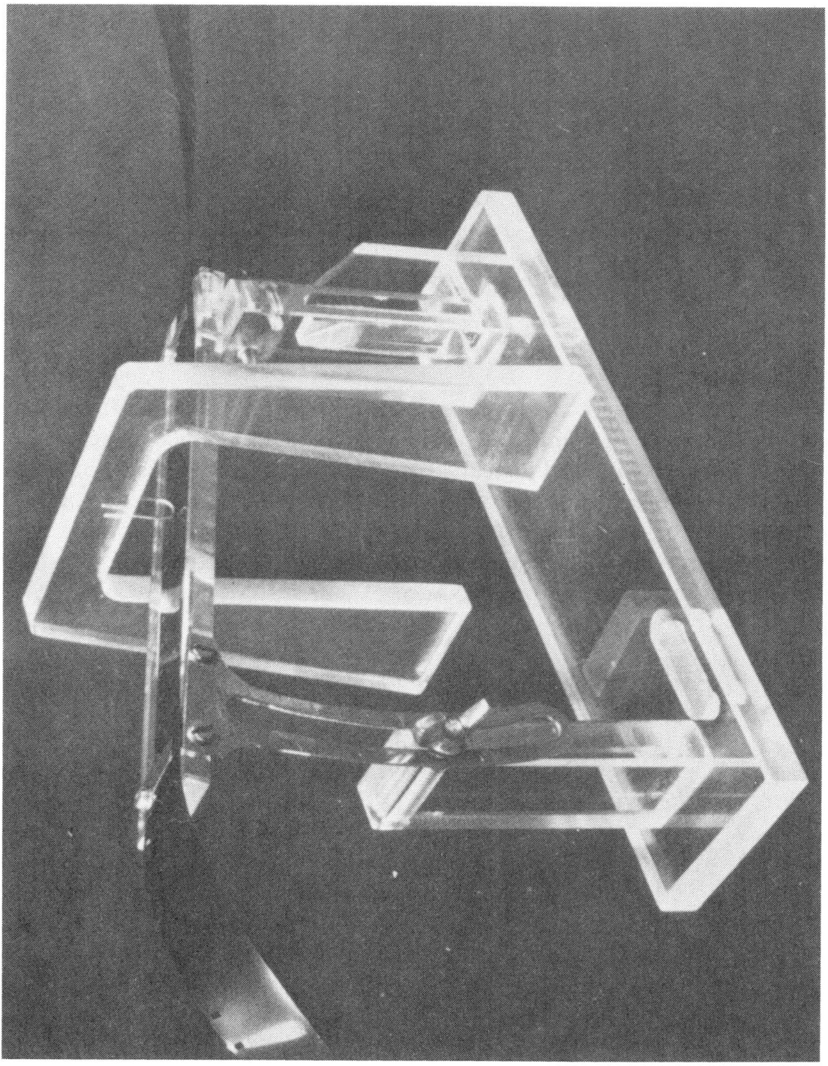

FIG. 13—*Paper clip friction tester.*

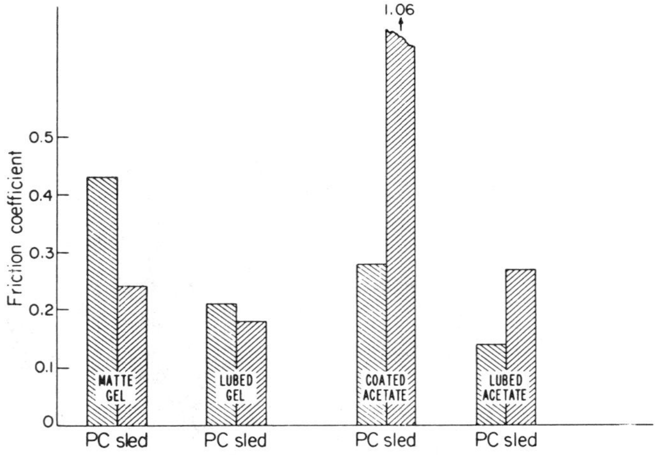

FIG. 14—*Paper clip versus sled friction.*

established after practical trade experience with photographic film. This method employs a high unit pressure over a small area. Once its pertinency was determined, it followed logically that a pertinent friction test should also have a high unit pressure. The Point Contact Friction Test was developed and has been used to measure and explain practical abrasion behavior. Likewise the high-unit-pressure Paper Clip Friction Test was a natural outgrowth to understand film projection performance.

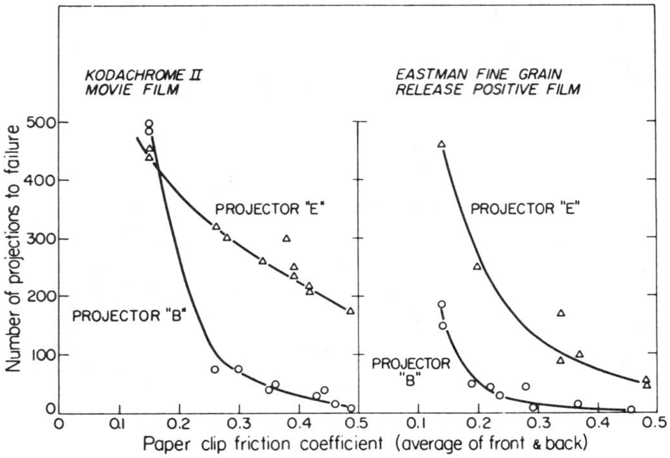

FIG. 15—*Effect of projector and friction on projection life 21.1° C (70° F) 50 percent relative humidity).*

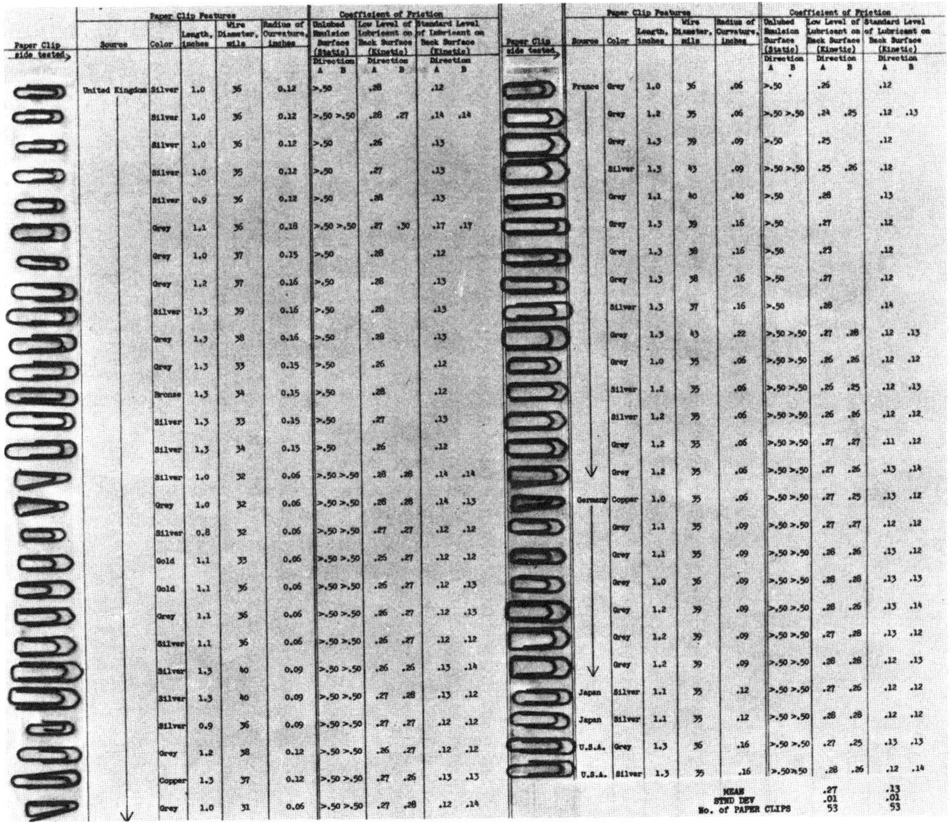

Source	Color	Length, inches	Wire Diameter, mils	Radius of Curvature, inches	Unlubed Emulsion Surface (Static) Direction A	Unlubed Emulsion Surface (Static) Direction B	Low Level of Lubricant on Back Surface (Kinetic) Direction A	Low Level of Lubricant on Back Surface (Kinetic) Direction B	Standard Level of Lubricant on Back Surface (Kinetic) Direction A	Standard Level of Lubricant on Back Surface (Kinetic) Direction B
United Kingdom	Silver	1.0	36	0.12	>.50		.28		.12	
	Silver	1.0	36	0.12	>.50	>.50	.28	.27	.14	.14
	Silver	1.0	36	0.12	>.50		.26		.13	
	Silver	1.0	35	0.12	>.50		.27		.13	
	Silver	0.9	36	0.12	>.50		.28		.13	
	Grey	1.1	36	0.18	>.50	>.50	.27	.30	.17	.17
	Grey	1.0	37	0.15	>.50		.28		.12	
	Grey	1.2	37	0.16	>.50		.28		.13	
	Silver	1.3	39	0.16	>.50		.28		.13	
	Grey	1.3	38	0.16	>.50		.28		.13	
	Grey	1.3	33	0.15	>.50		.26		.12	
	Bronze	1.3	34	0.15	>.50		.28		.12	
	Silver	1.3	33	0.15	>.50		.27		.13	
	Silver	1.3	34	0.15	>.50		.26		.12	
	Silver	1.0	32	0.06	>.50	>.50	.28	.28	.14	.14
	Grey	1.0	32	0.06	>.50	>.50	.28	.28	.14	.13
	Silver	0.8	32	0.06	>.50	>.50	.27	.27	.12	.12
	Gold	1.1	33	0.06	>.50	>.50	.26	.27	.12	.12
	Gold	1.1	36	0.06	>.50	>.50	.26	.27	.12	.13
	Grey	1.1	36	0.06	>.50	>.50	.26	.27	.12	.13
	Silver	1.1	36	0.06	>.50	>.50	.26	.27	.12	.12
	Silver	1.3	40	0.09	>.50	>.50	.26	.26	.13	.14
	Silver	1.3	40	0.09	>.50	>.50	.27	.28	.13	.12
	Silver	0.9	36	0.09	>.50	>.50	.27	.27	.12	.12
	Grey	1.2	38	0.12	>.50	>.50	.26	.27	.12	.12
	Copper	1.3	37	0.12	>.50	>.50	.27	.26	.13	.13
	Grey	1.0	31	0.06	>.50	>.50	.27	.28	.12	.14
France	Grey	1.0	36	.06	>.50		.26		.12	
	Grey	1.2	35	.06	>.50	>.50	.24	.25	.12	.13
	Grey	1.3	39	.09	>.50		.25		.12	
	Silver	1.3	43	.09	>.50	>.50	.25	.26	.12	
	Grey	1.1	40	.40	>.50		.28		.13	
	Grey	1.3	39	.16	>.50		.27		.12	
	Grey	1.3	38	.16	>.50		.23		.12	
	Grey	1.3	38	.16	>.50		.27		.12	
	Silver	1.3	37	.16	>.50		.28		.14	
	Grey	1.3	43	.22	>.50	>.50	.27	.28	.12	.13
	Grey	1.0	35	.06	>.50	>.50	.26	.26	.12	.12
	Silver	1.2	35	.06	>.50	>.50	.26	.25	.12	.13
	Silver	1.2	35	.06	>.50	>.50	.26	.26	.12	.12
	Grey	1.2	35	.06	>.50	>.50	.27	.27	.11	.12
	Grey	1.2	35	.06	>.50	>.50	.27	.26	.13	.14
Germany	Copper	1.0	35	.06	>.50	>.50	.27	.25	.13	.12
	Grey	1.1	35	.09	>.50	>.50	.27	.27	.12	.12
	Grey	1.1	35	.09	>.50	>.50	.26	.26	.13	.12
	Grey	1.0	36	.09	>.50	>.50	.28	.28	.13	.13
	Grey	1.2	39	.09	>.50	>.50	.28	.26	.13	.14
	Grey	1.2	39	.09	>.50	>.50	.27	.28	.13	.12
	Grey	1.2	39	.09	>.50	>.50	.28	.28	.12	.13
Japan	Silver	1.1	35	.12	>.50	>.50	.27	.26	.12	.12
Japan	Silver	1.1	35	.12	>.50	>.50	.28	.28	.12	.12
U.S.A.	Grey	1.3	36	.16	>.50	>.50	.27	.25	.13	.13
U.S.A.	Silver	1.3	35	.16	>.50	>.50	.28	.26	.12	.14
MEAN							.27		.13	
STND DEV							.01		.01	
No. of PAPER CLIPS							53		53	

FIG. 16—*Paper clip friction evaluation.*

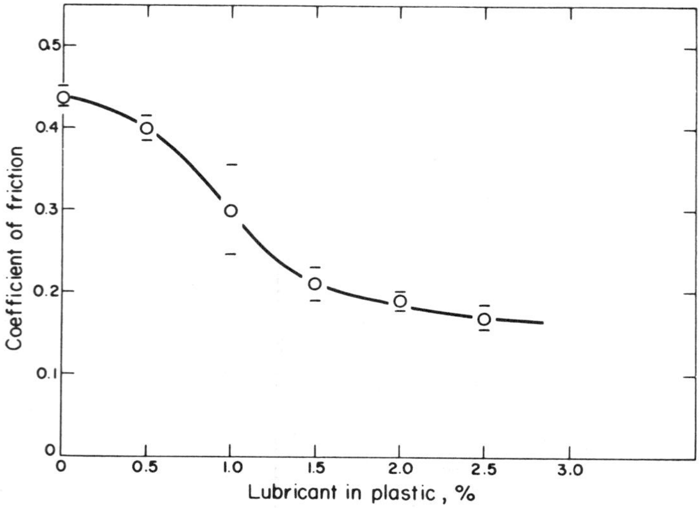

FIG. 17—*Paper clip friction versus nominal percent lube in molded plastic.*

Conclusion

Existing standard test methods for surface properties, albeit very precise, may not relate meaningfully to the practical application of a given end use. Your need for a more relevant test method today could be the beginning of a new relevant ASTM standard test method tomorrow.

Acknowledgments

Acknowledgment is made to P. Z. Adelstein for his advice and many valuable suggestions. Thanks are also due to J. L. McCrea and D. F. Wasielewski who furnished some of the experimental results.

References

[*1*] Carroll, J. F., Paul, J. O., *Photographic Science and Engineering*, Vol. 5, No. 5, Sept.-Oct. 1961, pp. 288-296.

[*2*] Talbot, R. H., *Journal of the Society of Motion Picture and Television Engineers*, Vol. 53, Sept. 1949, pp. 285-292.

[*3*] Johnson, E. C., *Journal of the Society of Motion Picture and Television Engineers*, Vol. 67, Sept. 1958, pp. 590-595.

[*4*] Kolb, F. J., and Weigel, E. M., *Journal of the Society of Motion Picture and Television Engineers*, Vol. 74, April 1965, pp. 297-307.

[*5*] Anvelt, T., Carroll, J. F., Sugden, L. J., *Journal of the Society of Motion Picture and Television Engineers*, Vol. 80, No. 9, Sept. 1971, pp. 734-739.

[*6*] Rabinowicz, E., *Friction and Wear of Materials*, Wiley, New York, 1955, p. 62.

N. S. Eiss, Jr.[1]

Wear Testing—Correlation with End-Use Performance

REFERENCE: Eiss, N. S., Jr., "**Wear Testing—Correlation with End-Use Performance,**" *Physical Testing of Plastics—Correlation with End-Use Performance, ASTM STP 736,* R. E. Evans, Ed., American Society for Testing and Materials, 1981, pp. 91–100.

ABSTRACT: The failure modes of a plastic involved in sliding motion are presented. The mechanisms of loss of plastic material by wear are discussed. These mechanisms include adhesive wear, in which the transfer-back transfer process occurs, abrasive wear in which loose debris particles are produced directly from the sliding action, fatigue wear which is often the mechanism by which transferred films wear, and degradation wear which results from interface temperatures high enough to decompose the plastic. Common tests, such as the abrasive wear test (ASTM D 1242-69) and the Pressure Velocity (PV) test are described, and the limitations in correlation with end use are noted. Suggestions are made which will improve the correlation between tests and end use.

KEY WORDS: plastics, abrasive wear, adhesive wear, fatigue wear, degradation wear, wear tests

The ever increasing use of plastics at sliding contacts in engineering systems has resulted in increased interest in understanding the mechanisms of wear of plastics. The wear mechanisms can be grouped in three classes: deformation wear, which includes abrasive and fatigue wear, interfacial wear, which includes adhesive wear and material transfer, and chemical wear, which includes material loss by melting or degradation. While these major wear mechanisms have been identified, there are no wear models which permit the designer to predict wear from measured properties of the materials in contact, the known interactions between the materials, lubricants, and the atmosphere, and the operating conditions. Consequently, the designer is forced to use accelerated tests to select materials which will have a satisfactory wear life in sliding contact.

In this paper, wear will first be described in terms of its relation to the engineering system. The basic wear mechanisms will be described, and some significant correlations with plastic properties and surface topography will be noted. Two common tests will be described and related to the wear mechanisms. Finally, some suggestions will be made for selecting tests for plastics wear which have high probability of predicting end use performance.

[1]Professor of Mechanical Engineering, Virginia Polytechnic Institute and State University, Blacksburg, Va. 24061.

Wear as a System Output

The application of the systems concept to the study of wear is one attempt to increase the awareness of designers and engineers of the complexity of the wear process [*1*].[2] As the wear process becomes better understood, the designers will be able to select tests which will better correlate with end-use performance. In this section, a plastic bearing, an example of a simple engineering system, will be discussed in a generalized description of the engineering system.

The engineering system consists of the plastic bearing and the shaft. The purpose of this system is to transmit motion (rotation, translation, or both), and torques and forces from one position to another in a more complex system. In addition, the bearing must constrain the position of the shaft axis to a specified tolerance. The inputs to the system and the useful outputs are the transmitted motions, torques, and forces.

However, all systems have two additional outputs which are usually undesirable, wear and friction. For the plastic bearing, the friction increases the input torque required for a desired output torque. The frictional heat also causes the properties of the plastic to change, altering the dimensions of the bearing. The wear of the bearing causes the shaft axis to change its position and eventually move out of tolerance.

In this example, the system consists of a stationary element (the bearing), a moving element (the shaft), possibly a lubricant, and the atmosphere. The properties of each of these system components, geometric, structural, physical, chemical, and thermal, interact to determine the outputs for a given set of inputs. For example, moisture in the atmosphere could corrode the metal shaft and absorb into the bearing, causing geometric changes of the surfaces of the shaft and the plastic bearing. The surface roughness features of the shaft could abrade the plastic, causing a rapid change in the position of the shaft axis. (See Ref 2 for a comprehensive discussion of dry bearings).

To generalize this example, wear is one of the outputs of a system subjected to a set of input conditions. The system consists of its elements, their properties, and their interrelationships. The system elements include the stationary and moving elements, the lubricant, and the atmosphere. The major properties of the elements are geometrical, structural, mechanical, and thermal. The interrelationships include oxidation between the atmosphere, the stationary or moving elements, or both. Deformation, adhesion, fatigue, and abrasion occur between the stationary and moving elements. The inputs and useful outputs of the system consist of motion transfer, information transfer, mechanical energy transmission, and materials transport. In most systems, wear and friction are considered as loss outputs. (Friction in brakes and wear caused by coated abrasive products are examples of friction and wear as useful system outputs.)

[2] The italic numbers in brackets refer to the list of references appended to this paper.

At present, the knowledge about the interactions between system elements is insufficient to permit the designer to predict wear from measured plastic properties, element geometry, and inputs. However, the basic wear mechanisms are known and have been correlated to certain polymer properties and geometries. These correlations are discussed next.

Wear Mechanisms

The wear of polymers is a complex phenomenon which is often simplified by assuming that one mechanism dominates the wear process. For example, when a polymer slides on a rough hard surface where the asperities are sharp, abrasive wear dominates. If the asperities on the surface are rounded so that the stresses in the polymer are elastic, fatigue wear dominates. If the surface is smooth, ductile polymers will transfer to the surface and form films. When the pressure at the area of contact and sliding velocity causes enough heat to be generated at the interface to raise the polymer temperature to its melting point or to cause degradation, high wear results [*3*]. Of these mechanisms, abrasive wear and melting and degradation cause the greatest loss of material per unit time.

The simplest wear mechanism is single-traversal abrasive wear, in which the sliding polymer is always being exposed to new abrasive surface. This wear mechanism is not complicated by the modification of the surface by polymer transferred on the previous traversal. However, Lancaster [*4*] has shown that in single-traverse abrasive wear, the material transferred from the leading edge of the slider influences the transfer from the trailing edge. Thus, for sliders with rectangular apparent areas of contact, higher wear was measured where the sliding direction was parallel to the short side, than where the sliding was parallel to the long side.

Single traversal abrasive wear has been found to correlate with a variety of polymer properties and surface topography parameters. Positive correlations have been found between wear and the inverse of the product of stress and elongation at rupture [*5-7*], the cohesive energy [*8*], and flexural modulus. There are inverse correlations to the yield strain [*9*] and molecular weight and crystallinity [*10*]. Polymer wear has been also found to correlate positively with the following measures of the topography of the abrading surface: arithmetic average roughness R_a, asperity slopes [*6,11*], and the ratio of the standard deviation of asperity heights to the average radius of the asperities [*12*]. Polymer properties and surface topography have been combined in a model which has predicted wear over a limited set of conditions [*13*].

Multiple traversal abrasive wear is characterized by an initially high wear rate which decreases to a steady state value as the number of traversals increase. This wear behavior is shown schematically in Fig. 1. Lancaster [*2*] noted that a minimum of two wear measurements were needed to determine the steady-state wear. As noted in Fig. 1, if wear measurements were only

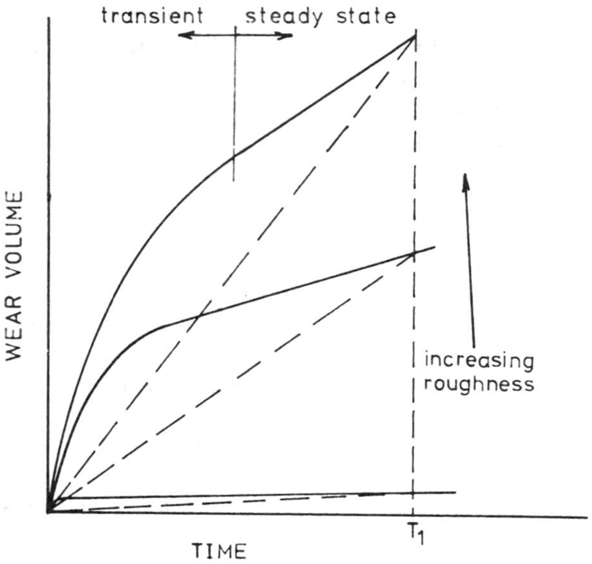

FIG. 1—*Wear versus time for multi-traversal abrasive wear.*

made at time T_1, the wear rate calculated would be higher than the steady-state wear rate, and lower than the initial wear rate.

The interrelationships that govern the steady-state wear are not understood as well as those that control single traversal wear. One example of the interrelationships between the geometric properties of the abrading surface and the plastic properties is the influence that the lay of the abrading surface has on wear [*14*]. Wear rate is lowest when the sliding direction is parallel to the lay, highest when sliding at an angle to the lay, and medium when sliding perpendicular to the lay. Consequently, single traversal abrasion tests are recommended because the results are more reproducible [*15*].

Wear rate decreases as the surface roughness decreases. Some investigators have found a value of roughness at which wear rate was a minimum. The roughness value varied from a high of 0.38 μm rms [*16*] to 0.03 μm R_a [*17*]. One reason given for the increase in wear rate at smaller roughness values is a change in mechanism from abrasive wear to interfacial wear. For ductile plastics, the interfacial wear takes the form of film formation.

For most ductile polymers the films formed are 0.1 to 1.0 μm thick. However, high density polyethylene (HDPE) and polytetrafluoroethylene (PTFE) transfer in films 10 to 50 nm thick at low sliding speeds and intermediate temperatures [*18*]. For the thicker film transfer, the film thickness is limited by removal of the film by fatigue and delamination [*19*], or by the elastic energy stored in the film exceeding the adhesion energy holding the film to the surface [*20*].

At high loads and sliding speeds the frictional heating causes plastics to melt and the wear rate to increase two orders of magnitude [*21,22*]. In some experiments, evidence of gas evolution has been used to infer the temperatures at the interface [*23*].

While the mechanisms of wear have been identified, and wear has been correlated with polymer properties and surface geometry, there are no models designers can use to predict wear rates. They must rely on measurements made in wear tests.

Wear Tests for Plastics

Laboratory wear tests for plastics are designed to provide a comparison between candidate materials, quickly and at minimum cost. Unfortunately, the laboratory tests provide no information on the wear rate which will be experienced when the material is put into application. At best, the material chosen from the laboratory test results will give the lowest wear rate of those materials considered for the application if the mechanism of wear is the same in the lab tests and the application.

In order to obtain test results quickly, high wear rates must occur in the laboratory tests. For this reason, many laboratory tests involve sliding the plastic against a bonded abrasive belt in single traversal sliding or using a plastic slider with a small cross section area in the sliding contact, high normal loads, and high sliding speeds. In addition, the sensitivity of the laboratory measurements are increased by using nonconforming geometries (ball sliding on a flat surface) rather than conforming surfaces (shaft in a plastic bearing). There is considerable controversy about the value of accelerated tests in predicting the wear behavior of materials in practical conditions. However, there can be reasonable correlation between laboratory tests and bearings as shown in Fig. 2 [*2*].

Because the abrasive wear mechanism causes high wear rates, many laboratory test machines have been developed using bonded or loose abrasives [*24*]. One of these machines shown in Fig. 3, is the basis for the ASTM Test for Resistance to Abrasion of Plastic Materials (1242-69). Single traversal sliding is designed into the test machine, thus providing as high a wear rate as possible with a given bonded abrasive belt, and improving reproducibility by eliminating the effects of transferred plastic on the transfer process in multiple traversal tests.

When a sliding system has reached a constant wear rate (steady-state wear), numerous experiments have shown that wear is proportional to the sliding distance. If it is also assumed that changes in load do not cause any significant change in temperature, then the wear is also proportional to the load. Under

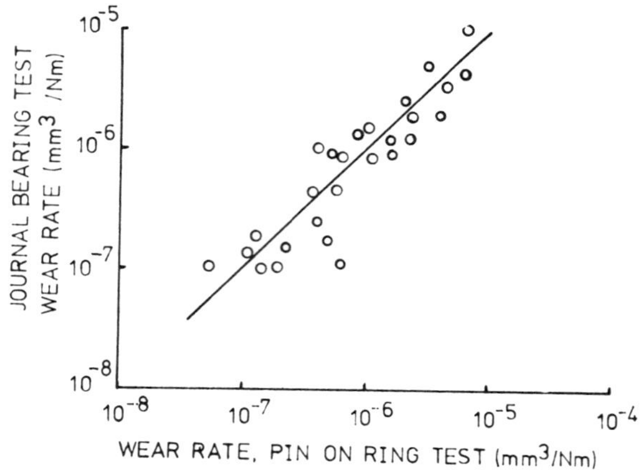

FIG. 2—*Correlation between pin and ring and journal bearing tests for carbon fiber reinforced polymers sliding on mild steel* [*2*].

these assumptions, the change in a linear dimension (h), per unit time (t), of a plastic due to wear can be expressed by

$$h/t = KPV$$

where
k = experimentally determined factor called the specific wear rate,
P = nominal applied pressure, and
V = sliding speed [*2*].

This equation leads directly to the concept of a *PV* factor as a performance criterion. The *PV* factor may take two forms:

1. A limiting *PV* above which wear increases rapidly.
2. A *PV* factor for continuous operation at an arbitrarily selected wear rate.

The limiting *PV* or *PV* factor is obtained by a pressure step test, a velocity step test, or a wear test [*25*]. The limiting *PV* is determined from either of the first two tests by holding one parameter constant and stepping up the value of the other until the wear increases rapidly. For values of *PV* below the limit, the wear per unit time is measured at steady-state conditions. The combined *PV* limit and wear results can be plotted on a single graph as shown in Fig. 4. One wear machine recommended for these tests is the thrust-washer tester in which the area of contact remains constant as the plastic washer wears [*25*].

The *PV* limit is usually not valid at low speed and high load because the strength of the plastic is exceeded, and wear proceeds more rapidly. At low loads and high speeds, the frictional heating can cause melting or rapid softening, again causing the wear to proceed more rapidly than predicted by the *PV* limit. The concept of the *PV* limit is associated with the softening of the plastic due to the increasing energy input and the resulting increase in wear rate as noted in the previous section.

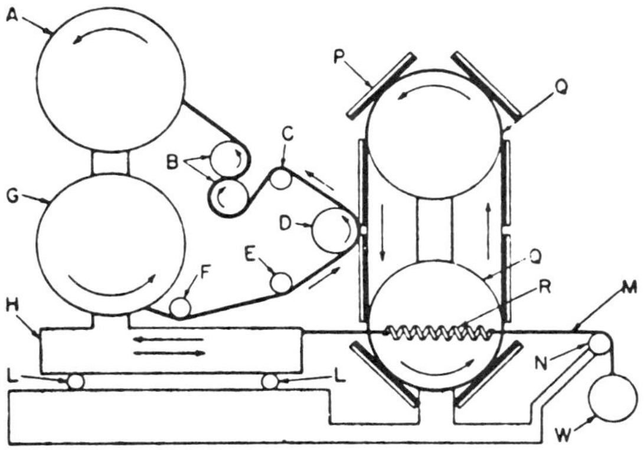

A.–*Take-up drum*
B–*Constant-speed-driven rolls*
C–*Abrasive tape*
D–*Steel contact roll*
E–*Slotted guide roll*
F–*Slotted guide roll*
G–*Abrasive tape supply drum*
H–*Carriage*
L–*Roller bearings*
M–*Spring cable*
N–*Pulley*
P–*Specimen mounting plates*
Q–*Constant-speed-driven pulleys with continuous link belt*
R–*Spring*
W–*Dead weight*

FIG. 3—*Schematic drawing of bonded abrasive abrading machine* (*D 1242-69*).

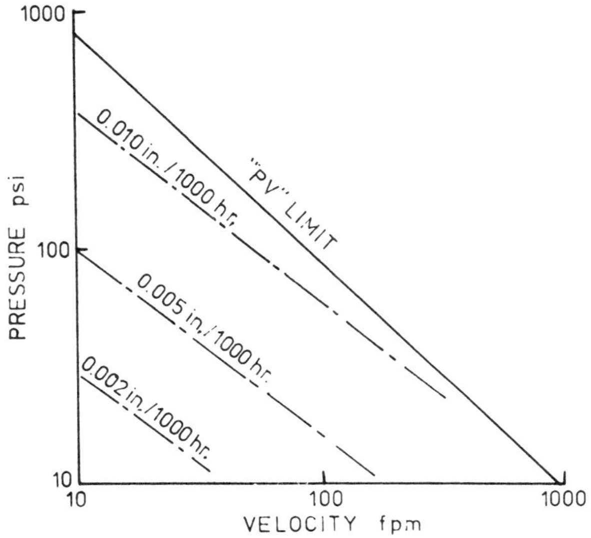

FIG. 4—*Wear and* PV *limit.*

Discussion

There is no argument that the "best" test to select a plastic for its wear resistance is to compare candidates in the actual end use. "Best" is used in the sense that the test will identify the candidate with the highest wear resistance, and also provide a quantitative measure of the wear rate and life. However, this test will usually not be the quickest test, and consequently will not be the least expensive. It is these latter considerations that have led to the use of accelerated tests for evaluation of plastics.

For applications where abrasive particles like dust may accidentally enter the sliding contact and cause catastrophic failure, resistance to abrasive wear would be the major factor in the materials selection. Hence, D 1242-69 would give a valid ranking of the materials for this application. However, the wear rates in the test and the application may be quite different because of the difference in the size, shape, and hardness of the abrasive particles present in each.

If the probability of abrasive particles entering the sliding interface is insignificant, then attention is shifted to the topography of the surface in contact with the surface. Lancaster [*2*] noted that the reduction in plastic wear achieved by grinding and polishing surfaces to values of R_a less than 0.05 μm, is more than offset by the increased machining costs. In this range of surface roughness, interfacial wear begins to dominate the wear process. Interfacial wear is influenced by the adhesion between the plastic and the surface, the ductility of the plastics and its ability to form films, and on the thickness of films at steady state wear. Consequently, a test such as D 1242-69 would not be expected either to rank plastics correctly or to predict the wear rates in the applications.

The *PV* limit or *PV* factors described in the previous section is not a test. For a given test or application, the *PV* limit indicates the upper bound of the product of pressure and velocity for steady-state wear. The *PV* factor indicates the measured wear rate at a specific value of input energy (or *PV* value). Thus, it is important to note that the *PV* limit and wear measurements are not a material property but are characteristic of the test (or engineering system). Hence, the *PV* limit measured in an accelerated test must be used with extreme caution in predicting its limit in an application. The more closely the test duplicates the application, the more accurate the prediction of the *PV* limit.

Are additional standardized tests needed? In an earlier section, three groups of wear mechanisms were identified as deformation wear, interfacial wear, and chemical wear. The abrasive wear test such as D 1242-69 is sensitive to the resistance of plastics to abrasive wear. The *PV* limit, while not a test apparatus, does outline a procedure for identifying the input energy at which the high wear rates due to chemical wear occur.

However, standard tests do not exist which measure (1) wear due to fatigue (plastic rubbing on a rough surface with blunt asperities), (2) the ability of a plastic to form a film on a smooth surface, or (3) the thickness of films found on a surface and their wear rates. It thus appears that additional standardized tests are necessary to provide the designer with measures of all of the major identified wear mechanisms of plastics.

As was noted above, D 1242-69 is based on the simplest of wear mechanisms, single pass abrasive wear. The development of standardized tests for other wear mechanisms will be difficult because of the randomness of the wear processes. However, the exercise in developing a standard will result in the generation of information for the designer, which may end up being more valuable than the existence of a standard test.

Conclusions

Of the three major categories of wear mechanisms which have been identified: deformation wear, interfacial wear, and chemical wear, only standard test D 1242-69 for abrasive wear is available for the designer to use in selecting plastics. The *PV* limit is a procedure which identifies the energy input at which the high wear rates due to chemical wear dominate the wear process. No tests are available to measure fatigue wear or interfacial wear phenomena. A positive correlation between laboratory tests for wear and end use requires that the dominant wear mechanism in both be the same, and that major elements and their properties in the test system and the application system be as closely matched as possible.

References

[*1*] Czichos, H., *Tribology, A Systems Approach,* Elsevier, Amsterdam, 1978, p. 29.
[*2*] Lancaster, J. K., *Tribology,* Vol. 6, 1973, p. 219.
[*3*] Briscoe, B. and Tabor, D., "The Sliding Wear of Polymers, a Brief Review" in *Fundamentals of Tribology,* N. P. Suh and N. Saka, Eds., Massachusetts Institute of Technology Press, Cambridge, Mass.
[*4*] Lancaster, J. K., *Transaction American Society of Mechanical Engineers, Journal of Lubrication Technology,* Vol. 97, 1975, p. 187.
[*5*] Ratner, S. B., Farberova, I. I., Radyerkevich, O. V., and Lur'e, E. G., *Abrasion of Rubber,* D. S. James, Ed., MacLaren, London, 1967, p. 145.
[*6*] Lancaster, J., *Wear,* Vol. 14, 1969, p. 233.
[*7*] Warren, J. H. and Eiss, N. S., *Transaction of the American Society of Mechanical Engineers,* Vol. 100, 1978, p. 92.
[*8*] Giltrow, J. P., *Wear,* Vol. 15, 1970, p. 71.
[*9*] Lontz, J. F. and Kumnick, M. C., *Transactions of the American Society of Lubrication Engineers,* Vol. 16, 1973, p. 276.
[*10*] Deanin, R. D. and Patel, L. B., *Advances in Polymer Friction and Wear,* L. H. Lee, Ed., Plenum Press, N.Y., 1974, p. 569.
[*11*] Eiss, N. S. and Warren, J. H., "The Effects of Surface Finish on the Friction and Wear of PCTFE Plastic on Mild Steel," Society of Manufacturing Engineers Paper No. IQ75-125.
[*12*] Hollander, A. E. and Lancaster, J. K., *Wear,* Vol. 25, 1973, p. 155.
[*13*] Eiss, N. S., Jr., Wood, K. C., Smyth, K. A., and Herold, J. H., *Transaction of American Society for Mechanical Engineers, Journal of Lubrication Technology,* Vol. 101, 1979, p. 212.
[*14*] Eiss, N. S., Warren, J. H., and Quinn, T. F. J. in *Wear of Non-Metallic Materials,* D. Dowson, M. Godet, and C. M. Taylor, Eds., Mechanical Engineering Publications Ltd., London, 1976, p. 18.

[*15*] Ratner, S. B. and Farberova, I. I., *Abrasion of Rubber*, D. S. James, Ed., MacLaren, London, 1967, p. 297.
[*16*] Buckley, D. H., *Advances in Polymer Friction and Wear*, L. H. Lee, Ed., Plenum Press, N.Y., 1974, p. 601.
[*17*] Dowson, D., Challen, J. M., Holmes, K., and Atkinson, J. R., *The Wear of Non-Metallic Materials*, D. Dowson, M. Godet, and C. M. Taylor, Eds., Mechanical Engineering Publications Ltd., London, 1976, p. 99.
[*18*] Pooley, C. M. and Tabor, D., *Proceedings of the Royal Society of London*, Vol. A329, 1972, p. 251.
[*19*] Suh, N. P., *Wear*, Vol. 25, 1973, p. 111.
[*20*] Rabinowicz, E., *Friction and Wear of Materials*, Wiley, N.Y., 1965, p. 151.
[*21*] Tanaka, K. and Uchiyama, Y. *Advances in Polymer Friction and Wear*, L. H. Lee, Ed., Plenum Press, N.Y. 1974, p. 499.
[*22*] Kar, M. K. and Bahadur, S., *Wear of Materials, 1977*, W. A. Glaeser, K. C. Ludema, and S. K. Rhee, Eds., American Society of Mechanical Engineers, N.Y., 1977, p. 501.
[*23*] Rhee, S. H. and Ludema, K. C., in *The Wear of Non-Metallic Materials*, D. Dowson, M. Godet, and C. M. Taylor, Eds. Mechanical Engineering Publications Ltd., London, 1976, p. 11.
[*24*] Gavan, F. M., *Testing of Polymers*, J. V. Schmitz, and W. E. Brown, Eds., Interscience Publishers, N.Y., Vol. 3, 1967, p. 139.
[*25*] Lewis, R. B., *Testing of Polymers*, J. V. Schmitz, and W. E. Brown, Eds., Interscience Publishers, N.Y., Vol. 3, 1967, p. 203.

F. S. Pang[1] *and J. L. Isaacs*[1]

The Need for More Descriptive and Meaningful Physical Data on Plastic Materials

REFERENCE: Pang, F. S. and Isaacs, J. L., "**The Need for More Descriptive and Meaningful Physical Data on Plastic Materials,**" *Physical Testing of Plastics—Correlation with End-Use Performance, ASTM STP 736,* R. E. Evans, Ed., American Society for Testing and Materials, 1981, pp. 101–108.

ABSTRACT: During the past few years, plastics have become the major structural materials for most of the new small electrical appliance products. Thus, when the decision is made to develop a new product, a plastics material screening process will be followed, where a list of material performance requirements (processing included) will be generated, based on the expected product performance, and then matched with currently available commercial grades of plastics.

Unfortunately the plastic material physical data furnished by the supplier are not adequate. They are either too general, use wrong sample preparation method, or do not give a meaningful description of the plastic material properties from processing to end-product performance. The new General Electric plastic iron and drip coffeemaker products are two typical examples which are used to illustrate that there is a definite need for making available to the end user more descriptive and meaningful physical performance data.

KEY WORDS: plastic materials, physical properties, design and performance requirements, small electrical appliance, coffeemaker, plastic iron

The small electrical appliance business (housewares) is traditionally a fast-moving, highly competitive, extremely innovative industry, where a product's demand in the marketplace is very difficult to predict. The General Electric Company has remained for many years one of the dominant forces in this industry through product diversification and by application of a responsive business strategy. As far as product diversification is concerned, the product line includes almost all food preparation, personal care, garment care, and home security appliances. All of these products can be divided into two categories; the "core products," or long-life, profitable business, and the short-life, fad, or "traffic appliances," with unpredictable return on investment.

[1] Project manager and manager-resident engineering, respectively, General Electric Company, Bridgeport, Conn. 06602.

In order to better understand these categories, it is necessary to explain the small appliance product life cycle. Market research indicates that most new small appliance product markets peak about three to four years after introduction, or about 18 months after the general public recognizes the feature of the product. After the new product market peaks, it could either become a long-life "core product," or just simply fade away to become one of the many fad products no longer produced. Therefore, if a small appliance manufacturer wants to be an important factor in the marketplace, particularly for the short-life products, he must recognize a new product concept early, develop the product quickly, provide easily recognizable consumer benefit, and assure dependable, safe performance. Research also indicates that a typical diversified small-appliance company will experience about 20 percent product attrition each year, that 20 percent of total product line is either new or has undergone a major redesign, and that the rest of the product line has usually gone through minor redesigns to improve quality, performance or appearance.

Plastic Materials in Small Appliances

To all of us at this conference, this means that small appliance design engineers must react quickly, since such a short product development cycle exists, and thus we must depend on material suppliers to furnish useful and meaningful material information. We have very little time to develop the property data and to conduct material performance testing for the intended product application, so we must depend on the information available to us at the time we need it and in a form we can understand.

As plastic materials continue to make significant inroads into small appliance applications, the need for more descriptive and meaningful physical data has become even more evident. In order to provide a better understanding of this point, we would like to take you through part of the new product development sequence we normally follow.

Once the decision is made for the development of a new product, a list of expected material performance requirements, which includes physical properties as well as processability, is generated. These requirements are then matched with currently available commercial grades of plastics. This is where problems have developed. We have found that vendor data sheets are usually presented simply as promotional information, leaving out many critical properties and not showing the material's deficiencies. At best, we have been able to use them only as a guide. We have found that almost every aspect of the physical property data (processability, as well as product consistency) must be checked in one way or another by either our own laboratory, sometimes with our own test methods, or through outside sources.

To illustrate this point, we would like to use a relatively new product, the Drip Coffeemaker's Water Reservoir, as an example (Fig. 1). The very first step in this program was to develop the list of material requirements based on the input from marketing, industry design, engineering, and manufacturing. The list might resemble the following:

1. Must be Federal Drug Administration (FDA) approved—both resin and pigment.
2. Must be transparent at room temperature as well as 100° C (212° F) for at least ten years of normal use.
3. Ultra Violet (UV) resistance.
4. Heat resistance—up to 100° C (212° F) as molded.
5. Adequate impact of molded part when assembled (temperature range) −28.9° C (−20° F) to 100° C (212° F).
6. Oil and food chemical resistance, including staining and different types of water found in United States.
7. Hot water and steam resistance.
8. No odor and taste problem.
9. Must have high modulus both at room temperature and 100° C (212° F) (Load bearing—16.7N (6 oz).
10. Easy to process—high volume production/automation.
11. Lowest cost possible.
12. Long-term availability, more than single supplier.

To illustrate the dilemma we face using the available technical data, let's take some of these requirements and match them with the typical vendor data based on either the ASTM test methods or vendor's own.

Meet FDA Requirements for Food Contact with Color-Matched Plastics

Vendor data do not mention what colors might have problems meeting FDA requirement. Furthermore, there is seldom any information concerning the FDA status of the base polymer.

Transparency at Room Temperature and at Approximately 100° C

The only information available is the light transmission rate from the Test for Haze and Luminous Transmittance of Transparent Plastics (ASTM D1003-61) which uses 23° C ± 2 only. There is no information about plastic clarity at either low (below zero) or high (above boiling, at near maximum) temperature range. What about the long-term clarity in this abnormal environment? What about the inflow/cross flow clarity effect, moisture effect (prior and post molding)? In almost all cases, the limited data available are all obtained under ideal, controlled conditions. We must test the plastics for all other requirements in our own laboratories.

UV/Sunlight Resistance

If the plastic might be affected by UV or sunlight, what are the common effects? How much time will it take before any changes occur? Very little information is available in this area. The ASTM Practice for Outdoor Weathering of Plastics, (D 1435-75) does cover a lot of ground. However, plastic suppliers seldom publish these data and usually never bother to check them.

Heat Resistance

Most of the data available are based on compression-molded specimens, annealed injection molding specimens, or compression-molded and annealed specimens. These data, in reality, do not apply to normal injection-molded parts. Furthermore, the DTUL data obtained by following ASTM Test for Deflection Temperature of Plastics under Flexural Load (D 648-78) only indicate a single directional load, with no mention of plastic flow direction and molded-in stress during specimen preparation. As an example of what the plastic DTUL tells us and what the facts actually showed, the published data on a particular formulation SAN had a DTUL of 112°C (235°F) at 1264 Pa (264 psi), with no mention of whether specimens were molded or annealed. Based on the design calculation, this material should work for the coffeemaker water reservoir application, which will see only 98.9°C (210°F) maximum with no more than a 120 Pa (25 psi) external load, excluding molded-in stress. The actual test units could not last more than 12 cycles of operation before they began to sag (Fig. 2).

FIG. 1—*Drip coffeemaker reservoir.*

Processability of Plastic Material

Next, we would like to discuss the question of processability of the plastic material. As far as we know, there is no readily available information concerning wall thickness versus flow distance at preferred melt/mold temperature for most materials. The general guidelines do not apply to different molecular weights of the same polymer, the presence of filler or modifier, and many other factors. What about the possibility of material degradation under prolonged heat history while being trapped in the machine barrel or hot runner? What are the time/temperature relationships before degradation occurs? What are the common indications and property effects when degradation occurs?

The question we are raising could go on and on, but we think the point is made. The plastic physical property data necessary for the engineers and designers to make meaningful decisions do not exist or are not conveniently available. Therefore we, the engineers from the product manufacturers, are forced to depend on judgement, run controlled experiments, or develop our own data. All of this extra work leads to program delays and forces us into unnecessary testing. Our point is that we are not looking for tailored data which apply to our specific products, we are simply asking the vendors to supply more complete and meaningful data, and, in most cases, we are asking that they make available only the data they have within their files.

In order to further amplify the points we have raised, we would like to look at an innovative product which we have recently introduced into the Garment Care area, the Light'N Easy Iron (Fig. 3).[2] During the early stage of this development, our marketing personnel defined that they wanted an iron that must be cool to touch on sides (except the aluminum soleplate), that looks somewhat smaller, is about 50 percent lighter weight yet must have full features, come in attractive colors, and must meet our manufacturing and cost targets. These "wants" then were translated into a set of design requirements which told us that the plastic material must be able to withstand the expected heat and chemical environments, be somewhat flame resistant to meet Underwriters Laboratories and safety standards, be compatible with current manufacturing capabilities, process easily, and be low cost. The final design ended up with two major plastic parts. They are the thermoplastic top housing and the thermoset lower plate, or skirt.

Let us look at these parts as examples to illustrate our points about the missing, as well as misleading, vendor data.

1. The top housing uses a thermoplastic-polypropylene material. In all the published physical property data concerning polypropylene, no matter what filler or reinforcements are used, it is indicated that the maximum DTUL is about 130°C (266°F) at 316 Pa (66 psi) and 80°C (176°F) at 1264 Pa (264 psi) (ASTM D 64878). However, through use of a special formulation for a

[2]Trademark of the General Electric Company.

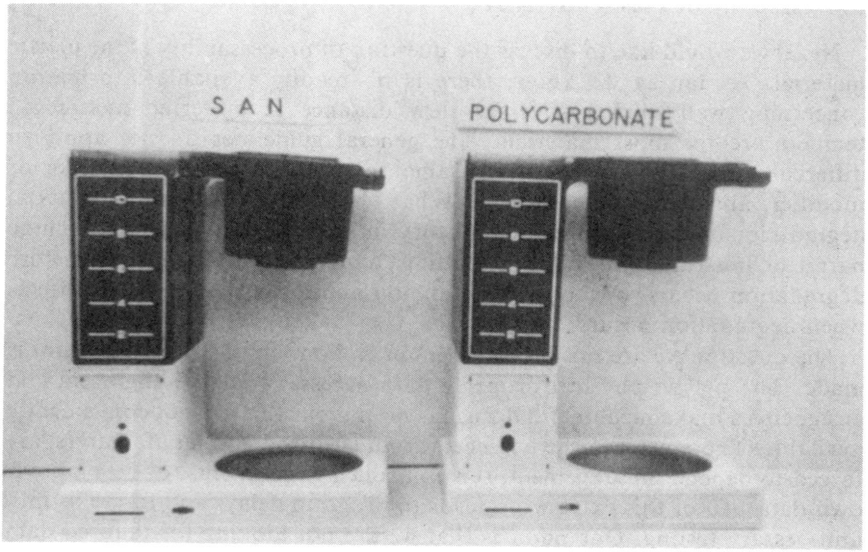

FIG. 2—*SAN reservoir and polycarbonate reservoir.*

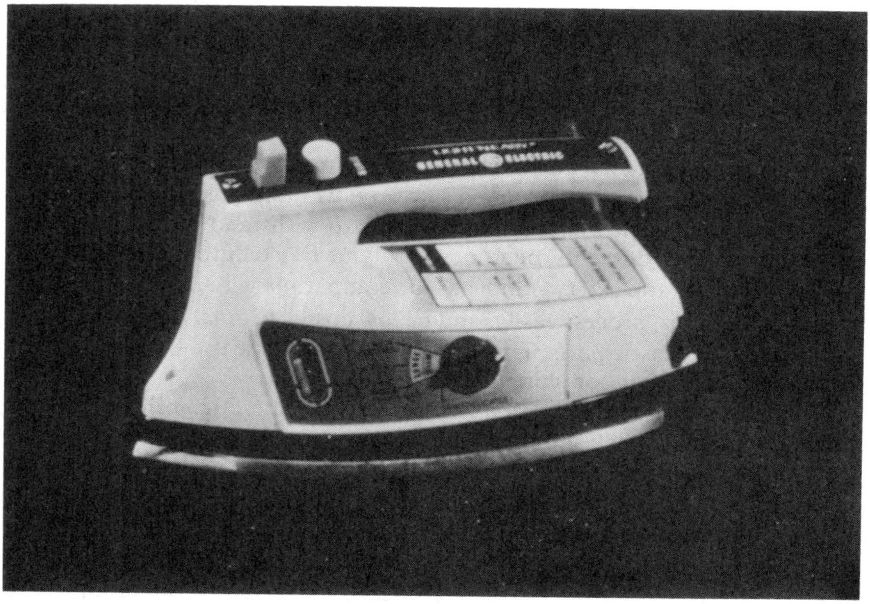

FIG. 3—*Light N' Easy iron.*

General Electric proprietary filled polypropylene and some innovative design, we are able to meet the UL 3000 hours continuous heat aging test at 147°C (295°F). In fact, our internal destructive test goes to an ever higher temperature, 160°C (320°F), for a minimum of 720 h. In this case, the vendor data give no indication that, through innovative part design, this family of materials can be successfully used up to 50 percent over its DTUL.

2. For the lower skirt plate, a special grade of reinforced phenolic compound is used (Fig. 4). If we based our design on the published data for this type of material, we would not have found the proper material. For example, the impact data always are based on 12.7-mm (0.5 in.) thick specimen, while we are designing this part with wall thickness of no more than 2.29-mm (0.090 in.). The product specification also requires a 0.9-m (3-ft.) drop on to an oak floor without significant breakage, which would allow a 6.35-mm (0.25-in.) probe to touch any exposed internal electrical component. If we just use the vendor's impact data, there are probably at least a dozen reinforced phenolic materials which will meet the impact requirement. However, evaluation with a 2.286-mm (0.09-in.) wall narrows the choice to only two. Because of this problem, we had to develop an in-house test apparatus to impact test specimen plaques as well as the molded part at approximately the same cross section (Fig. 5). Using this test method, we can easily qualify any potential candidate material without going through the costly product evaluation.

We are making this point because the industry's published data do not show the differences between these two materials and the other eight to ten

FIG. 4—*Light N' Easy iron skirt.*

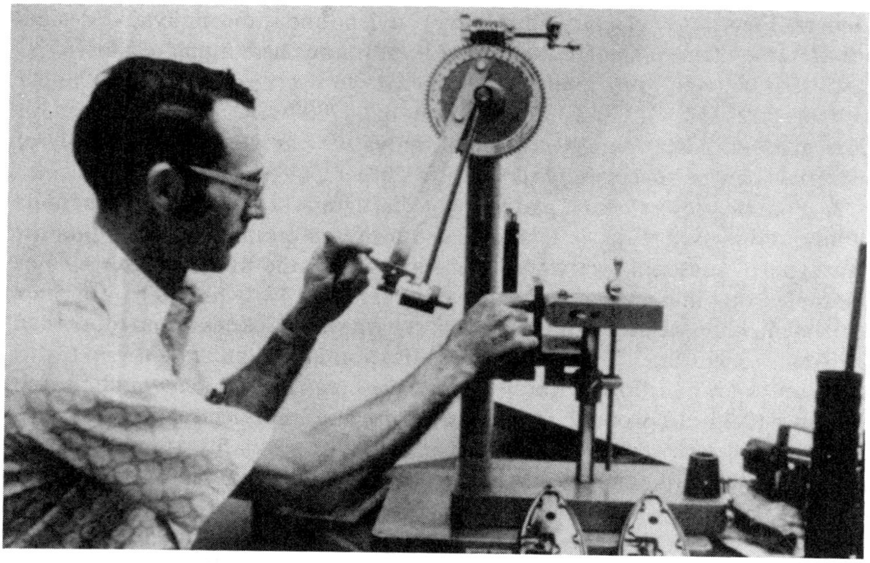

FIG. 5—*Iron skirt impact testing machine.*

candidates tried. Yet, by using a different test method at a more realistic material thickness, 2.286 mm (0.090 in.) two materials stand out. The plastic industry should look at the more realistic test methods and specimen thicknesses, closer to what the manufacturers use today on their products.

Conclusions

In conclusion, let us re-emphasize our position with regard to the problem we are facing. We are asking the materials suppliers for:

1. Complete data.
2. Realistic data.
3. Data presented in a form we can use.
4. Available data, when and where we need it.

The plastic industry has come a long way. Today the available physical data are certainly more useful and complete in many cases than in years past. No doubt these available data in the hands of experienced engineers serve as important references and bases for candidate material selection. But as you can see from these examples, we need more.

We connected with the end-user industries, call upon you, the ASTM committees, to take the lead in providing the more descriptive and meaningful physical data needed.

J. L. O'Toole[1]

Testing Plastics for Performance: Meeting the Challenge of the Future

REFERENCE: O'Toole, J. L., "**Testing Plastics for Performance: Meeting the Challenge of the Future,**" *Physical Testing of Plastics—Correlation with End-Use Performance, ASTM STP 736,* R. E. Evans, Ed., American Society for Testing and Materials, 1981, pp. 109–113.

ABSTRACT: Previous papers have analyzed the problems of testing viscoelastic materials for end-use performance, summarized the current state of the art, and suggested various approaches to improving our rate of progress. One proposition on which all agree, there is a great deal of work to do that is challenging both technically and logistically. One of the principal areas of challenge one whose solution would have far reaching benefits, is also one for which the American Society for Testing and Materials Committee D–20 (ASTM) is uniquely positioned to attack. This is the interface between viscoelastic material science and the technology of analytical design. More than any other single factor, the inherent interdisciplinary nature of plastics design has frustrated the development and promulgation of a complete and coherent performance testing system. A program and organization whereby D–20 can meet this challenge will be outlined.

KEY WORDS: design properties, performance tests, viscoelastic design, design systems, engineering properties

The growth of plastics during the past 40 years has been phenomenol by any standard. Much of this growth has been in traditional, relatively low stress applications, such as packaging film, detergent bottles, toys, throwaways, and household items. Today, however, the fastest growing applications area for plastics is in engineered parts, where performance under a variety of mechanical stresses and environmental conditions is critical, controlling material selection. This growth is due partly to the awareness by virtually all manufacturing industries of the cost and performance advantages of plastics, and partly to their potential for more efficient use of energy and feedstocks in many applications compared to structural metals and other traditional materials.

To take advantage of this growth opportunity fully, plastics must overcome a serious obstacle that has persisted in spite of its growth, namely, the lack of a coherent plastic design and strength-of-materials technology comparable in effectiveness to that of metals. To accomplish this, ASTM D–20 is in a unique position to play a key role.

[1]Industry manager, Allied Chemical Co., Morristown, N.J. 07960.

Application Trends

A number of application trends that grow out of both the versatility of plastics and the needs of industry are today producing greater emphasis for virtually all plastics on design, and particularly on the use of performance properties to select materials and to calculate dimensions. The most important of these are the following:

1. *Increasing part complexity* which takes advantage of versatile plastics, part fabrication techniques, and results in multi-functional parts that must bear multiple stresses including static loads, fatigue, impact, friction, and wear.
2. *Higher stresses* as plastics are applied more frequently in structural parts previously restricted to metals.
3. *Higher temperatures,* particularly in electrical and electronic devices and in machinery.
4. *Larger parts* taking advantage of economical plastics fabricating methods, such as blow molding, rotational molding, thermoforming, sheet molding, and filament winding. Inevitably larger parts tend to produce higher and more complex stresses.

The Problem of a Plastics Design Technology

The lack of an effective design technology for plastics can be viewed as a problem with four roots:

Viscoelastic Strength-of-Materials Is Different From Metals

Because plastics are viscoelastic, their performance under load is profoundly different from that of elastic structural metals. These differences require a significantly different approach to strength-of-materials testing and the engineering and design properties derived therefrom. For example, stress-strain properties, such as Young's modulus and yield stress which are widely used key design constants for metals, do not have the same significance for plastics. The stress-strain properties of plastics are not constants, varying greatly and nonlinearly with speed of loading and temperature. Therefore, they are inaccurate predictors of performance, often grossly inaccurate. Even at room temperature and low loads, creep and creep rupture must be usually considered in plastics design under static load, whereas for structural metals these are important only at very high temperatures. In virtually every type of stress involved in the design of parts, including fatigue, impact, friction, wear, snap fits, screw threads, interference fits, dimensional stability, etc., viscoelasticity causes significant differences in the determination of performance properties of plastics.

Lack of a Viscoelastic Design Technology

The only system of stress analysis available today is based wholly on the theory of elasticity, and uses material properties that depend on elastic behavior for validity. Unfortunately, for reasons of phenomenological complexity, a system based on the theory of viscoelasticity has not yet been developed and is not likely to be in the foreseeable future. Therefore, ways must be found for adapting viscoelastic properties to elastic design formulae.

Deficiency of Viscoelastic Course Material in Pertinent Engineering Curricula

For a number of reasons (some of which are historical and some inertial), engineers who bear the principal burden of design, mechanical, industrial and civil, are given woefully inadequate academic training in viscoelasticity, and therefore, tend to perpetuate the problem when they enter the work force.

Although there are many post graduate and seminar programs emphasizing chemical and material science aspects of plastics, there are few, if any, directed to the specific needs of the designer. No significant change in this situation is likely in the foreseeable future.

Misunderstanding and Misapplication of Existing ASTM Tests

Although a great deal is known of the viscoelastic performance of plastics, and much of this knowledge has been published in the chemical literature and textbooks on polymer science, little of this information has been translated into handbook technology that is usable by engineers responsible for part design and material selection. Certainly no standardized body of such technology has yet been developed. Designers have depended largely on their own experience. Consequently, most plastics design, to date, has been accomplished by costly and time consuming trial and error.

The properties of plastics that have been most available to designers are the traditional metal design properties, derived from stress-strain tests, plus data from arbitrary ASTM tests whose proper field of use is in material purchase specifications.

For the most part, these data-sheet properties have only a very limited or very general relationship to performance. The result has been frequent misapplication of ASTM test data for lack of proper interpretation and education.

The Challenge

The development of a comprehensive design technology for plastics is, of course, a long-term cause that will involve industry, universities, and professional societies for years to come. However, an order-of-magnitude step should be taken now, which would be of incalculable practical benefit to users and suppliers of plastics. That step is the generation of plastics design practices and performance test procedures based on the pooling of the plentiful available knowledge of plastics behavior under the aegis of ASTM D-20. However, for this work to be practical and useful, it is essential that it include a strong and continuing input from a group that is not now represented to any appreciable degree in D–20, namely, the designers themselves. Their discipline, which is different from testing and material science, is absolutely necessary to define the detailed objectives.

D–20 has ample precedent for this undertaking. It was in this committee that the system of design for pressure rated plastic pipe was developed and institutionalized in a remarkable body of ASTM specifications, test methods, and recommended practices. Now, under the jurisdiction of ASTM F–17, this system has provided the essential technical foundation for a thriving and eminently successful business that would not have developed without it. Outside of pipe, a considerable amount of performance test development has already been done by D–20 in the form of methods for creep, creep rupture, fatigue, drop-weight impact, high-speed stress-strain, and others. However, much more remains to be done, especially in the area of application to design.

Proposed Program

A formal D–20 program should be set up with the following general objectives:

1. Adapt viscoelastic properties to elastic design formulae where possible.
2. Define other performance properties to measure what the designer says he needs.
3. Develop (or modify existing) tests for the defined design and performance properties.
4. Develop and publish practices and guidelines that are readily and directly usable by part designers and product engineers including those who lack in-depth understanding of plastics behavior.

Proposed Organization

To clearly distinguish between those ASTM test methods that produce design and performance properties and those that are useful only for material purchase specifications, a new subcommittee should be formed with sole and complete responsibility for design and performance technology. Responsibility for existing methods that fall into the design and performance classification, such as creep and fatigue, should be transferred to the new committee.

Initially, the Design and Performance Subcommittee could be subdivided as follows, with revision and expansion as experience and resources dictate:

Section 1—Static Loading covering strength and rigidity under short and long term loading, prediction and extrapolation, recovery and set, heat resistance under load (short and long term), chemical resistance under load, and environmental stress-cracking.

Section 2—Shock Loading covering energy absorption without failure, effects of geometry, and effects of stress mode.

Section 3—Cyclic Loading covering endurance limit, hysteretic heating, and effects of stress mode.

Section 4—Dimensional Stability.

Section 5—Effects of Fabrication Variables.

Conclusion

ASTM D–20 is in a unique position to render a vital and badly needed service to industry because it possesses the machinery for integrating material technology, test methods, and design practices into discrete elements of a system of plastics design and performance technology and, also because it can publish the results as self-educating, national standards. To accomplish this, D–20 must first, secure the working input of designers and product engineers, and second, organize its effort to clearly distinguish design and performance tests from purchase specification tests.

Summary

The purpose of this volume has been to discuss physical testing required for the successful assessment of a material or article in relation to the function it is to fulfill, that is, to establish suitability for a given application. Unique philosophy and methodology have been shown to be required for this, compared to the other reasons for testing such as quality assurance, obtaining standard reference or typical value data, or research to obtain a fundamental understanding of polymer structure. The philosophy of physical testing presented by Evans has been echoed in various forms throughout the remainder of this book. The words "relevancy of the test" are stated or implied in every paper.

Physical testing was defined, in phenomenological terms, as the measurement of the output response of a polymer system under given boundary conditions and in a given surrounding environment when it is subjected to external input physical excitations. It is shown that the influence of the time-form of the excitation functions, the boundary conditions and the surrounding environment on the nature of the output response must be considered in any test method selection. Unfortunately, many testing technologists have tended to overlook this in favor of assuming that physical properties of polymers can be represented as unchanging values that describe material behavior. Because of this, it is estimated that 70 percent of physical testing data is either useless or ignored.

Test methods employed in physical testing have been divided into three classes. Fundamental tests yield data concerning basic physical properties independent of the boundary conditions. Component tests yield data from a finished part to allow the prediction of article behavior under service conditions. Hybrid tests yield data from special specimen geometries or machines with results dependent on the boundary conditions. Practical tests, while dependent on boundary conditions, are modeled so that the boundary conditions of the test are identical with those to which the component would be exposed. Adoption of a sound philosophy of engineering measurement has been repeatedly emphasized requiring fundamental, component or practical tests relevant to end-use performance to be used whenever possible in place of typical value measurements.

As the first example where typical value data cannot be used to judge end-use performance concerned prediction of the bursting strength of polyethylene pipe containing flaws or scratches. For such cases, a fracture mechanics approach is adopted. Fracture toughness testing of plastics is in its infancy, especially in regard to the acceptance of standard procedures for its determination. However, it has been shown to be a critical aspect of many physical testing programs where correlation with end-use performance is expected. The fracture toughness of a material is a fundamental property analogous to strength or modulus. However, determination of this property requires that the sample fail by brittle fracture. Fuller, et al, examined the validity of the fracture mechanics adopted from metals technology. Single edge notched specimens cut directly from pipes were tested in three point bending. Three modes of failure were observed under varying conditions ranging from ductile to brittle. The authors show that the validity question is critical, since simple calculation of the fracture toughness is shown to give erroneous, nonconservative values if the fracture is not sufficiently brittle due to specimen size, test conditions, or precrack introduction procedure. By careful selection of conditions and fractographic analysis, it is shown that it is possible to apply fracture mechanics to high density polyethylene. The agreement between the predicted and experimental pipe burst data for HDPE pipe with longitudinal external cracks indicated that standard procedures used for metals may serve as a guideline when the test parameters are adjusted to obtain the correct output response, brittle fracture for example. Data presented show that predictions of performance assuming only that the crack reduces wall thickness predict gross overestimates of burst strength. However, because of the ductile nature of polymers, the general validity of fracture toughness equations to particular applications should be closely examined. Nevertheless, this paper illustrates that fracture mechanics techniques can be employed that correlate with actual end-use performance. Several ASTM task groups are presently concerned with the general application of fracture mechanics to polymeric materials.

A second example where relevant physical evaluations were required concerned measurements of the long term behavior of plastics under either sustained dynamic or static loads.

Two excellent examples of meaningful fatigue testing were discussed by Loveless. These examples also clearly demonstrate the rationale to be followed when evaluating a material for a known application. Fatigue can be considered as a fundamental test. However, it is necessary that the shape and size of the input excitation have the same form encountered in the end-use application. In the first application, flexural fatigue tests, adopted from ASTM Test for Flexural Fatigue of Plastics by Constant-Amplitude of Force (D-671), using tensile-compressive load input were able to predict the secondary bonding characteristics of resins employed in the fabrication of boats. Those tests simulated the input excitations encountered by the boat in choppy waters. Such tests were adopted after several other ASTM standards for short term properties including flexural, tensile and lap shear strengths gave inconclusive, often meaningless results.

The other example concerned static tensile fatigue encountered in the steel core of aluminum conductor, steel-reinforced cable used in electrical power transmission. Since such cables are expected to have a normal life expectancy of 35 to 40 years, the long term tensile strength of the load carrying core is the fundamental property that should be measured. Since the short term tensile strength of glass reinforced polyester rod is 82 percent of steel, it is suggested that a designer might be tempted to substitute a slightly larger glass reinforced rod for a steel rod to save weight. However, when the long term strengths were predicted from accelerated tests, the glass reinforced rod had 50 percent of the strength of steel. Also, it is shown that the lower stiffness glass reinforced core prevents it from becoming the principle load-bearing component as is the case for steel. This analysis prevented misapplication of a plastic that could have proved disastrous resulting in a disservice to both the end-user and the plastics industry.

Recently developed impact testing equipment is discussed in the next paper which has permitted a shift from the classical hybrid types of impact testing such as Izod, Charpy, and Gardner to practical or component types of tests that can directly predict performance. Classical tests are based on either the measurement of total absorbed energy or by "go no-go" evaluations. The boundary conditions of these types of tests have been stated as having little relation to those imposed in-service, a fact often ignored by the test technologist. With instrumented impact testers, it is shown that a complete record of the load applied to the specimen during the entire impact event is provided, in addition to the conventional single value of the total absorbed energy. Impact performance can be evaluated in terms of all characteristics of the fracture process. Thus, loads for fracture initiation can be measured as well as total failure. Also, the boundary conditions can be selected to correlate with the end-use application.

Ireland presents several examples of full penetration types of data which correlate with end-use performance and discusses load initiation and total energy required for rupture. Automotive flex fans, protective head gear, chain saw housings, etc. were some examples of tests with an instrumented apparatus. Another unique feature of instrumented impact tests has been shown to be the ability to conduct low blow types of tests, where the impacter does not completely penetrate the specimen, that is, it delivers just enough energy to cause incipient damage. Such measurements of the extent of damage are necessary for example to assess the damage caused by a tool dropped on an epoxy/graphite wing section of an airplane. Incipient damage can be detected from the change in slope of the initial elastic loading.

The use of instrumented drop impact testing for the general evaluation of a material and the establishment of bounds for acceptable performance is discussed by Rieke. The mode and velocity of impact is stated to simulate actual blows that a material such as PVC siding would be expected to experience in service. The types of failure mechanisms and the ductile-brittle transition for six formulations of PVC showed wide differences even though the room temperature data indicated that most materials would perform equally well. The importance of including an environmental aging study were also emphasized. The relative behavior instead of the absolute values as a function of temperature permitted the ranking of materials and the selection of the most suitable candidate. The critical recognition in this paper is that when general evaluation of materials is being made using other than fundamental tests, only relative rankings of materials can be made because the absolute values change as the specimen geometry are changed. The presentation of these data in a graphical form also make them very useful for the designer to determine the bounds for ductile behavior.

Instrumented impact tests were also shown to be able to easily assess the effect of fabrication as shown by the different behavior of two injection molded refrigerator trays. This illustrates the need to fabricate the test piece in the same manner that the material will be fabricated in end-use. The effect of thickness on impact strength is another parameter that is shown to be easily assessed and the data capable of being used in a design for end-use. Linear relationships between thickness and energy to failure were not observed even though many standard methods still recommend normalizing impact data. Thus, meaningful impact tests should be conducted using test specimens having the same thickness as the intended part designed for end-use.

Both the discussions by Ireland and Rieke illustrate that instrumented impact measurements can provide more information for the same amount of effort and material than can be obtained by the more classical methods being utilized. The interest in instrumented impact testing is increasing and ASTM is considering the development of standard methods.

The next two papers discuss various surface tests and the special techniques that are recommended for their correlation with end-use performance. Surface testing is particularly difficult because by definition such tests are dependent on the boundary conditions, or mating surfaces, and are thus either hybrid or practical. This has been clearly indicated by Carroll in searching for the most appropriate abrasion tests to evaluate photographic film. Falling carborundum particles and horsehair brushes did not produce abrasion patterns that correlated with end-use performance. This is not surprising since both tests were hybrid. When a spherical stylus scratch test was used, which produces an abrasion by the same type of high unit pressure over a small area common to film sliding or feeding through various camera and developing mechanisms, excellent correlation with actual performance was found.

The same rationale applies to explain why a point contact friction test was more sensitive to assessing the effects of film lubrication than the standard ASTM Test for Static and Kinetic Coefficients of Friction of Plastic Film and Sheeting (D-1894-78) flat sled. The test also allowed the role of lubricity in changing scratch resistance to be quickly determined.

Once the relevancy of a surface test for an end-use application has been demonstrated, the actual test apparatus might be very simple. This has been illustrated by the paper clip friction tester which was designed to detect film lubricants. A close relationship exists between the projection life of photographic films and their level of lubrication as quantified by this test.

Predictions of wear performance has been shown to be another area where laboratory tests are generally hybrid in nature and considerable insight is required to establish that the boundary conditions of the test are related to end-use. Eiss uses the concept of an output response to an input excitation to study wear. This system approach requires that the interaction of the material, geometric, structural, physical, chemical, and thermal components be considered.

The wear mechanisms identified involve deformation wear which includes abrasive and fatigue wear, interfacial wear which includes adhesive wear and material transfer, and chemical wear which includes material loss by melting or degradation. While the mechanisms of wear have been identified, there are no models that the designers can use to predict wear rates. They must rely on measurements made in wear tests.

Unfortunately, as Eiss notes, most wear tests are designed with an eye on accelerated tests at a minimum cost. They usually provide no information on the wear rate which will be experienced when the material is used in end-use. The standard ASTM Test for Resistance of Plastic Materials to Abrasion (D1242-75) for example, should only be expected to correlate with performance when abrasive wear dominates. The pressure-velocity (PV) limit should correlate when chemical wear dominates. The establishment of generally applicable wear tests is doubtful. Rather, the designer should analyze each situation and develop practical tests which match the boundary conditions with those of end-use. However, in trying to develop new test methods, the author recognizes that a great deal of information may be generated for the designer that will produce better correlations between test results and end-use performance.

The paper by Pang and Isaacs discusses the designers impressions of ASTM testing procedures for establishing suitability for an end-use application. Attention is called to the large amount of nonessential data for product design that is presented in product sheets. As found throughout this volume, the need for fundamental properties as a function of temperature, time under loading and under other environmental influences is emphasized. The problems encountered in the development of a plastic iron and drip coffeemaker are used as two examples to illustrate that there is a definite need for making available to the end-user more descriptive and meaningful physical performance data. Many of these problems can be related to the failure of the supplier to provide fundamental properties of the material over a wide range of conditions. For example, had a fundamental thermal mechanical profile obtained by a technique such as dynamic mechanical analysis been presented instead of a single misleading hybrid heat distortion temperature, the high temperature performance of a polypropylene material would have been obvious. The authors note that better presentations of physical properties would have aided the selection of materials.

In the last paper, O'Toole discusses the need for a coherent plastic design and strength of materials technology comparable in effectiveness with that of metals. Because of the viscoelastic behavior of plastics, however, characterization of their performance requires techniques different from that used for metals. To overcome the special problems encountered with plastics, O'Toole recommends the generation of plastics design practices and performance test procedures based on the pooling of available knowledge of plastics behavior using both the disciplines available in ASTM D–20 as well as those that would need to be recruited, that is, the designers themselves. To accomplish this a sweeping reorganization of Committee D–20 is proposed. The proposed reorganization, however, is more extensive than required to start development of design and performance criteria. This paper does, however, present an excellent summary of the kind of philosophy that ASTM Committee D–20 must adopt if it is to meet the challenge of the future. Fundamental, practical, and component test methods needed for design and performance must be clearly distinguished from purchase specification tests.

Unfortunately, some topics presented at the symposium are not included in this volume. One of the most important areas is mechanical spectrometry. A simple thermal mechanical profile of the energy stored, (modulus) and energy dissipated (Tan δ) by a plastic perhaps yields more insight and fundamental information about such viscoelastic materials than any other single measurement. The reader is urged to consult other texts to find demonstrations of the contributions of dynamic mechanical measurements to end-use applications.

Another area that is not discussed in this volume concerns the use of portable microprocessors to collect field data, strain and acceleration values for example, on an actual component in service so that the magnitude and shape of input excitations given to the laboratory specimens can be adjusted to be relevant with the end-use conditions.

The problems of testing for meaningful predictions of end-use performance have been summarized, the current state of the art reviewed, and various approaches suggested for improving our rate of progress. One proposition on which there is general consensus is that there is a great deal of work to do that is challenging both technically and logistically.

What then are the priorities for improved physical testing techniques which better correlate with end-use performance in the future? On the basis of the papers included in this volume, it appears that a rationale and methodology for meaningful testing can be established. However, such future development will require closer liaison between the testing experimentalist and the designer. More than any other factor, the inherent interdisciplinary nature of plastics usage has frustrated the development and promulgation of a complete and coherent performance testing program.

As O'Toole has stated, ASTM Committee D–20 is in an excellent position to act as a coordinating focus in any efforts to develop relevant testing procedures. An initial step has been taken with the formation of Section D–20.10.24 on "Design and Performance Criteria." The goal of this D–20 section is to provide guidelines for the selection of procedures and physical properties applicable to the reliable design and performance of plastics. Hopefully designers and test engineers will use this common forum to express and exchange ideas leading to the development of new standards. Further progress in these areas should be reported regularly in future publications of this sort.

R. E. Evans
American Cyanamid Company, Stamford, Conn. 06904; symposium chairman and editor.

Index

A

Abrasion tests, 10, 77–85, 93
  Armstrong abrasion, 95, 97, 98
  Mar resistance, 78, 117
  Performance predictions (*see* Physical testing)
  Soft-hair wheelbrush, 79–80, 117
  Spherical stylus scratch, 79, 85
  Taber abrasion, 79
Abrasive wear (*see* Wear mechanisms)
ANSI test procedures, 80, 82, 85

B

Boundary conditions, 1, 7, 9, 10, 11, 13, 59, 114, 116, 118
  Definition, 6
  Geometry, 21, 95
  Thickness, 7, 9, 59, 72–73, 76, 107–108, 117
  Topography, 92, 93, 94, 98

C

Chemical wear (*see* Wear mechanisms)
Component tests
  Definition, 8, 114
  Examples, 11, 29–30, 57–58, 70, 72, 104, 106
Creep, 12, 110, 112

D

Data presentation, 13, 108, 118
  Typical property tables, 1, 13, 102, 103, 105, 111, 113, 114
Deformation wear (*see* Wear mechanisms)
Design requirements, 1, 103, 105, 110, 112, 117
  Design technology, 109, 110, 111, 118
  Role of ASTM, 112–113, 119
Ductile-brittle transitions, 22–29, 60, 64–70, 74, 116

E

Environments
  Effects of, 1, 3, 5, 7, 12, 16, 59, 63, 64–70, 76, 92, 95, 105, 114
  Inclusion in material evaluation, 13, 103, 104, 117, 118
Excitation functions (*see* Input/output functions)

F

Fabrication influences, 4, 7, 19, 22–29, 56, 59, 70–72, 102, 104, 105, 117
Fatigue testing, 32–44, 115
  Cyclic, 32, 35, 36
  Performance predictions (*see* Physical testing)
  Static, 32, 35, 38
    Prot method, 41, 42, 43
Fatigue wear (*see* Wear mechanisms)

FDA, 103
Fracture toughness, 16, 19–20, 30
  Crack opening displacement, 17, 18, 19, 20, 21
  Experimental methods, 16–17, 114, 115
  Performance predictions (*see* Physical testing)
  Stress intensity factor, 17, 21
  Validity conditions, 18, 20, 30
Friction testing, 80–85
  Paper clip friction test, 84, 85, 87, 88, 89, 90, 117
  Performance prediction (*see* Physical testing)
  Point contact friction test, 81–83, 84, 88, 117
  Sled friction test, 80, 84, 85, 117
Fundamental tests
  Definition, 7, 114
  Examples, 8, 10–11, 16, 41, 115, 118

H

Heat resistance (*see* Environments)
Hybrid tests
  Comparison with practical tests, 10
  Definition, 9–10, 114
  Examples, 10, 11, 13, 78, 79, 117, 118

I

Impact testing
  Charpy, 46, 51, 57, 116
  Falling weight, 53, 56, 57, 60, 61
  Gardner, 61, 116
  Instrumented, 46, 116
    Advantages, 45–46, 54
    Applications, 49, 53
    Incipient damage, 52–53, 116
    Instrumentation, 46–49, 60–61
  Izod, 7, 9, 10, 11, 12, 46, 57, 116
  Performance predictions (*see* Physical testing)
  Special impact tests, 107
Input/output functions, 1, 6–7, 11, 13, 92, 98, 114, 118
  Shape and amplitude, 6, 115, 119
Interfacial wear (*see* Wear mechanisms)

L

Load-deformation curves, 19–20, 47–49, 51–52, 54–56, 62–63

M

Material evaluation guidelines
  General evaluation, 13
  Known application, 12–13
Mechanical spectrometry, 7, 118, 119
Microprocessors, 119

O

Optical haze
  Light transmission, 103
  Measurement of abrasion, 78, 79, 80, 81

P

Performance requirements (*see* Design requirements)
Physical testing
  Correlation with end-use performance, 1, 4, 12–13, 29–30, 35, 43, 44, 50–53, 57–58, 78, 80, 81, 83–85, 88, 90, 91, 92, 95–96, 98–99, 104, 105–107, 114, 115, 116, 117, 118, 119
  Definition of, 6–7, 114
  Nonstandard methods
    Assessment, 79, 81, 84–85, 88, 107
  Rationale, 3, 119

Reasons for testing, 3–4, 77
Relevancy, 12, 77, 78, 85, 88, 91, 114, 115, 119
Standard methods
Assessment, 11–12, 30, 35, 53, 55, 60, 77–79, 80, 84, 90, 95, 98–99, 103–104, 107, 108, 111, 112, 115, 118
Comparison, 8–11
Types of methods, 7–10, 45, 114
Practical tests
Definition, 9, 114
Examples, 9, 11, 36, 45, 79, 81, 88, 98, 99, 107–108, 117
Processing (*see* Fabrication influences)

R

Rate effects, 21–22, 97, 98
Response functions (*see* Input/output functions)

S

Scratch tests (*see* Abrasion tests)
Stress analysis, 111
Stress rupture (*see* Fatigue, static)
Systems analysis (*see* Input/output functions)

T

Temperature (*see* Environment)
Temperature measurement, 61
Time
Effects of, 1, 3, 6
Inclusion in material evaluation, 13, 103
Long-term strength, 41, 44, 115–116

U

Underwriters Laboratories, 105, 107

V

Viscoelasticity 6, 110–111, 119
Strength of materials, 110
Stress-strain properties, 110

W

Wear
Mechanisms, 91, 92, 93, 94, 95, 98–99, 118
Performance predictions (*see* Physical testing)
PV factor, 96, 97, 98, 118
Specific wear rate, 96
Thrust-washer tester, 96
Wear testing, 91, 92, 95–97, 117
Wohler method, 41